25.00

Computers
in
Mass Spectrometry

Computers in Mass Spectrometry

J. R. CHAPMAN

*Kratos-AEI Scientific Apparatus Ltd.,
Urmston, Manchester, England.*

1978

ACADEMIC PRESS

*London, New York and San Francisco
A Subsidiary of Harcourt Brace Jovanovich, Publishers*

ACADEMIC PRESS INC. (LONDON) LTD
24–28 Oval Road,
London NW1

U.S. Edition published by
ACADEMIC PRESS INC.
111 Fifth Avenue,
New York, New York 10003

Copyright © 1978 By ACADEMIC PRESS INC. (LONDON) LTD

All Rights Reserved
No part of this book may be reproduced in any form by photostat, microfilm, or any other means without written permission from the publishers

Library of Congress Catalog Card Number: 77-92818
ISBN: 0-12-168750-3

Printed in Great Britain by
Page Bros (Norwich) Ltd., Norwich

Preface

Mass spectrometry and gas chromatography-mass spectrometry are no exception to the rule that modern analytical instrumentation produces large amounts of data at a prodigious rate. The fledgling liquid chromatography-mass spectrometry seems likely to follow where these techniques have led. As a result, computers are obligatory for the acquisition, handling and interpretation of the data produced.

The purpose of this book is not to describe in detail the instrumental hardware involved. Development in the computer field is too rapid to give such a text much credibility. Instead, I have attempted to describe the processes by which data are treated and the potential of the methods available.

The first chapter gives a brief introduction to the essential instrumentation—the mass spectrometer, the data system and the data acquisition interface. The book then follows the logical sequence of events in the use of this instrumentation, i.e. the acquisition of data, the conversion of the raw data to mass and intensity data and the reduction of this data to forms more intelligible and convenient to the user. Three subsequent chapters deal with the very wide ranging methods used to interpret the mass spectrometric data in analytical or chemical terms. The final chapter deals with the application of mass spectrometric data to quantitative analysis.

Mass spectra have provided a fertile field for the study of methods in computer handling and interpretation of data, and the 1970s have witnessed an explosion of publications in this area which has only just begun to subside. A practical assessment of these techniques by their prospective users is now due and I therefore make no apology for presenting a comprehensive comparison of these techniques in the hope of encouraging such efforts.

I am grateful to my friends and colleagues at Kratos-AEI Scientific Instruments for their help and advice in writing this book. My thanks are also due to Dr N. A. B. Gray for his criticism of Chapter 6 and to my wife for her help with the manuscript.

April, 1978   *J. R. CHAPMAN*

Contents

PREFACE — v

1. Instrumentation — 1

    1.1   Introduction — 1
    1.2   The mass spectrometer — 3
    1.2.1 Single-focusing magnetic sector instruments — 3
    1.2.2 Double-focusing magnetic sector instruments (Nier-Johnson geometry) — 5
    1.2.3 Double-focusing magnetic sector instruments (Mattauch-Herzog geometry) — 6
    1.2.4 Quadrupole mass spectrometers — 7
    1.2.5 Time of flight mass spectrometers — 8
    1.3   The mass spectrometer computer interface — 9
    1.4   The computer — 9
    1.4.1 The central processing unit — 9
    1.4.2 Data transfer — 11
    1.4.3 Core memory — 12
    1.4.4 Back-up memory — 13
    1.5   Input-output devices — 14
    1.6   Software — 15
    References — 17

2. Data Acquisition — 18

    2.1   Introduction — 18
    2.2   Data acquisition from scanning mass spectrometers — 18
    2.2.1 General methods — 18
    2.2.2 Electron multiplier and pre-amplifier — 21
    2.2.3 Analogue signal conditioning — 22
    2.2.4 Signal transmission — 23
    2.2.5 Signal multiplexing — 23
    2.2.6 Analogue to digital conversion — 23
    2.2.7 Interface logic and control functions — 29
    2.2.8 Digital thresholding — 30
    2.2.9 Peak recognition routines and data enhancement — 30
    2.2.10 Peak position and intensity calculation — 34
    2.3   Alternative data acquisition systems using an analogue to digital converter — 36
    2.4   Data acquisition from quadrupole mass spectrometers — 39
    2.5   Data acquisition from focal plane mass spectrometers — 40
    References — 44

CONTENTS

3. Data Conversion 46

    3.1   Introduction 46
    3.2   Nominal mass determination 46
    3.2.1 Calibration 46
    3.2.2 Determination of mass from peak times—rounding off to nominal mass 47
    3.2.3 Determination of mass using variables other than time . . . 51
    3.2.4 Instrumental stability 52
    3.3   Accurate mass measurement 52
    3.3.1 Calibration 53
    3.3.2 Determination of accurate masses by interpolation 55
    3.4   Errors of mass measurement 57
    3.4.1 Ion statistical errors 57
    3.4.2 Instrumental errors 59
    3.4.3 Errors of measurement and calculation 59
    3.4.4 Effect of digitization frequency on mass measurement accuracy . 60
    3.4.5 Effect of scan rate and resolving power on mass measurement accuracy 62
    3.4.6 Improvement of mass measurement accuracy 64
    3.5   Peak-matching methods 65
    3.6   Mass measurement from photoplates 66
    3.7   The recognition and processing of multiplet data 67
    3.8   The acquisition and processing of metastable ion data 69
    3.9   The acquisition and processing of ionization efficiency data . . 73
    References 74

4. Data Reduction 77

    4.1   Introduction 77
    4.2   Nominal mass data—scanning methods 77
    4.3   Nominal mass data—presentation 78
    4.4   Reduction and clean-up of repetitive scan nominal mass data— mixture analysis 79
    4.5   Elemental compositions from nominal mass data—calculation of monoisotopic spectra 91
    4.6   Reduction of accurate mass data—conversion to elemental compositions 95
    References 99

5. Library Search 101

    5.1   Introduction 101
    5.2   Data bases 103
    5.3   Abbreviation and encoding 106
    5.3.1 Encoding of a limited number of peaks 107
    5.3.2 Encoding as features 114
    5.3.3 Encoding as a mathematical function 116
    5.3.4 Use of intensity data in encoding 118

CONTENTS

5.4	Search and comparison routines	120
5.4.1	Logical operator comparisons	120
5.4.2	Comparisons based on intensity ranking	123
5.4.3	Reverse search systems using intensity ranking	125
5.4.4	Comparisons made using stored intensity data	125
5.4.5	Substructure searching using stored intensity data	131
5.4.6	Reverse search systems using stored intensity data	135
5.5	Filtering	139
5.6	Methods related to library searching	144
5.7	Summary	144
References		147

6. Pattern Recognition 150

6.1	Introduction	150
6.2	Classification by minimum distance from an average spectrum	151
6.2.1	Judging the performance of a classifier	155
6.2.2	Summary	
6.3	Classification by learning machines	157
6.3.1	Character of data—pre-processing and feature extraction methods	159
6.3.2	Choice of categories	162
6.3.3	Choice of training set	164
6.3.4	Linear separability	164
6.3.5	Multicategory prediction	166
6.3.6	Digital learning nets	167
6.3.7	Summary	170
6.4	Classification by K-nearest neighbour (KNN) method	170
6.4.1	Summary	174
6.5	Parametric classification methods	174
6.6	Unsupervised learning—cluster analysis	179
6.7	Classification using factor analysis	182
References		184

7. Spectrum Interpretation 187

7.1	Introduction	187
7.2	Interpretation of low resolution spectra	187
7.3	Peptide sequency	192
7.4	Interpretation of high resolution spectra	194
7.5	Heuristic dendral	199
7.5.1	Structure elucidation	199
7.5.2	Recognition of the molecular ion	204
7.5.3	Structure generation	205
7.5.4	Formalization of fragmentation rules	209
7.5.5	Summary	213
7.6	Use of metastable data in interpretative schemes	214
7.7	Summary	216
References		217

CONTENTS

8. Quantitative Analysis 220

    8.1   Introduction 220
    8.2   Quantitative data from scans of unresolved mixtures 220
    8.3   Quantitative data from repetitive scans of separated mixtures . . 223
    8.4   Stable isotope determination from scan data 225
    8.5   Selected ion monitoring 228
    8.6   Control of sector instruments for selected ion monitoring . . . 229
    8.7   Control of quadrupole instruments for selected ion monitoring . 235
    8.8   Data acquisition from selected ion monitoring experiments . . 236
    8.8.1 Use of analogue to digital converter systems 237
    8.8.2 Use of ion counting systems 238
    8.9   Data processing from selected ion monitoring experiments . . 240
    8.10   Quantitative data from spark source analyses 246
    References 251

Appendix 254
    Useful formulae and definitions 254

SUBJECT INDEX 259

1
Instrumentation

1.1 INTRODUCTION

Approximately ten years ago it was realized that the increasing amounts of experimental data that could be produced by a number of analytical methods necessitated the application of computer-based techniques. This problem was particularly acute in two areas of mass spectrometry, namely the acquisition of complete high resolution scans to give elemental compositions and the use of repetitive scanning techniques in gas chromatography-mass spectrometry (GCMS). The earliest solutions to these problems in mass spectrometry were realized by the use of off-line processing of recorded data. One approach involved the recording on analogue magnetic tape, of the output of the electron multiplier of a scanning mass spectrometer (MS) (1). This data would then be digitized off-line for subsequent computer processing. Alternatively, the output could be digitized directly and recorded on digital magnetic tape (2). Again, the recorded data would be subsequently processed by means of a computer. A similar approach was used with high resolution spectra recorded on photographic plates (3). These off-line methods were inherently time-consuming and very often the interposition of a data logging device, such as magnetic tape, imposed technical limitations on the system. Thus, the next logical step was the development of on-line systems whereby a direct link from the MS to the processing facilities of the computer was achieved and processed results were much more immediately available. A fortunate chance meant that the introduction of on-line systems coincided with the introduction on the market of reasonably priced minicomputers. The rapid expansion of mini-computer production and their continued ability to provide the increasingly sophisticated hardware required for on-line data processing systems at a reasonable price has allowed the introduction of these systems into a very large number of mass spectrometry laboratories.

Mini-computer systems for MSs are generally dedicated systems, that is a system in which the computer serves either one instrument or a number of instruments of the same type with uniform data processing requirements, e.g.

fast scanning MSs. Such dedicated systems are especially suited for fast scanning instruments because of their ability to deal with the high data throughput and also provide the best system for instrumental control and closed loop feedback operation.

In contrast to dedicated systems, installations that encompass a wide range of instrumentation can with advantage be constructed as time-sharing systems in which a large central computer can control the operation of thirty or more separate analytical instruments (4). Such a system can offer access to enormously powerful computing facilities for each analytical instrument and great flexibility in operation. For example, work on off-line programs (background processing) can be carried out concurrently with data acquisition from on-line instruments. Where one or more of the instruments is required to operate at the highest possible repetitive scan rate, these instruments can be interfaced to a smaller dedicated satellite computer that acquires data prior to transfer to the central time-shared computer. This is the basis of a hierarchical system (5). The hierarchical system also overcomes another disadvantage of the basic time-shared system, viz. it is tedious and expensive to change interface hardware.

Time-sharing and its eventual growth into a hierarchical system seems to be the preferred direction of progress for larger installations. However, such installations are of necessity expensive and the dedicated mini-computer system is, at present, the system of choice for average laboratories.

The most recent development in the computing field is that of the microcomputer (6, 7). The technology of large-scale integration (LSI) has made possible the manufacture of silicon chips doped with impurities to form electronic component patterns (resistors, transistors and diodes) necessary for specific functions. Thus, all the functions for a central processor unit (CPU) may be included on a single micro-processor chip. The combination of such a microprocessor with necessary memory elements and additional control elements results in a microcomputer.

Microcomputers are small so that they can be easily physically incorporated into analytical instruments, they are low in cost and consume little power. Thus, microcomputers are being incorporated into the MS itself as "intelligent terminals", replacing hardware control functions by much more flexible software. For example, a microcomputer-based interface could handle some real time control of the spectrometer and the conditioning and transmission of data in digital form to a central mini-computer which would effect the storage and processing of large amounts of recorded data. Conventional controls used to set instrumental parameters can be replaced by a microprocessor facility. Data systems designed for specific operating routines, e.g. isotope ratio determination (8), may also be microprocessor-based.

Another important development in computing is the growth of computer networks. The basis of a network is a large-scale computer which can be accessed by a very large number of interactive, time-sharing users. Access to the network is provided from the user's terminal by connecting this to a telephone line via an acoustic coupler or modem. Networks provide effectively unlimited computing facilities and storage as well as access to a large number of tested program packages. The major application of networks in mass spectrometry data processing so far, has been the installation and maintenance of large-scale library search systems, e.g. the Mass Spectral Search System (MSSS) available through CYPHERNET (9) and the Cornell STIRS (Self-training Interpretative and Retrieval System) and PBM (probability based matching) systems available through TYMNET (10). The Stanford Dendral programs (Section 7.5) are also available through computer network systems as part of the SUMEX-AIM Stanford University Medical Experimental Computer–Artificial Intelligence in Medicine project (11).

1.2 THE MASS SPECTROMETER

The basic components of a MS are shown schematically in Fig. 1.1. One of the most characteristic features of particular MSs is the method used to obtain mass separation of the characteristic ions produced from the sample

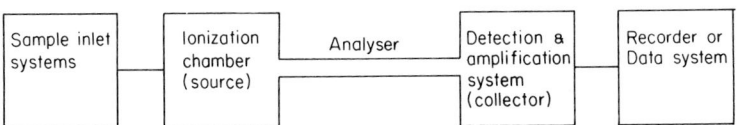

Fig. 1.1. Basic components of a MS.

in the ionization chamber, i.e. the mass analyser. The more important forms of mass analyser used in MSs are described below.

1.2.1 Single-focusing Magnetic Sector Instruments

The principle of the single-focusing magnetic deflection MS is illustrated in Fig. 1.2. Ions formed in the source are accelerated out of the source with a voltage V. For an ion to reach the collector slit and be recorded, it must traverse a path of radius of curvature r through the magnetic field. For an ion of mass to charge ratio (m/e) the equation of motion in a magnetic field H is:

$$\frac{m}{e} = \frac{H^2 r^2}{2V}. \quad (1.1)$$

Thus, by varying either H or V, ions of different m/e ratio can be made to reach the collector. In practice, scanning to record a complete mass spectrum

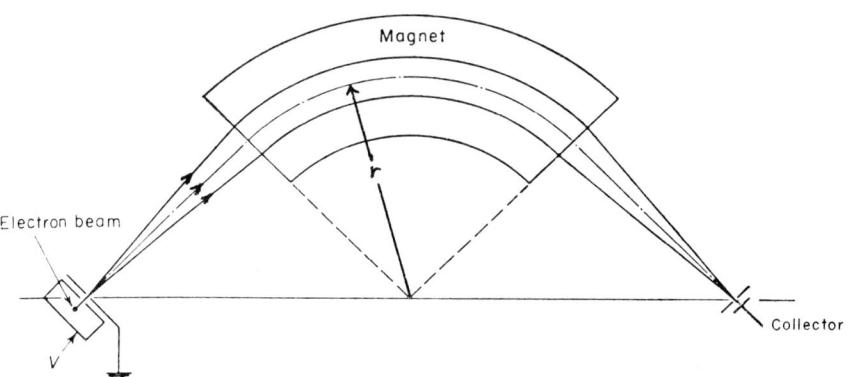

Fig. 1.2. Schematic of 90° magnetic sector showing direction focusing of divergent ion beam.

is usually accomplished by varying the magnetic field H. Scanning of the accelerating voltage V is used much less frequently because of mass discrimination problems. However, for applications in which selected compounds are to be monitored by successively monitoring only selected mass values, these mass values have commonly been brought to a focus at the collector by rapid stepping of the accelerating voltage V, whilst maintaining a constant magnetic field H. Such multiple peak monitoring (MPM) systems are discussed in more detail in Chapter 8.

The most common form of magnetic scan is the exponential, either upwards or downward in mass. It has the advantage of producing mass spectral peaks of constant width. The equations appropriate to this form of scan are:

$$m - m_0 c^{kt} \quad (1.2)$$

$$t_p = t_{10}/2 \cdot 303 \, R \quad (1.3)$$

where m_0 = starting mass at time $t = 0$; m is the mass registered at time t; t_p is the peak width between its 5% points; t_{10} is the time taken to scan one decade in mass (e.g. m/e 200 to 20 or m/e 500 to 50) and R is the resolving power measured on the 10% valley definition. R should be reasonably constant throughout the mass range. Numerous other forms of magnetic scan law are, of course, possible.

1.2.2 Double-focusing Magnetic Sector Instruments (Nier–Johnson Geometry)

Ions of the same mass but of differing energy are not brought to a point focus in a single-focusing magnetic deflection instrument. Thus, because of the energy distribution of ions formed in the source, only a limited resolving power can be obtained. This defect can be overcome by the addition of an electric sector which then provides a system with energy focusing.

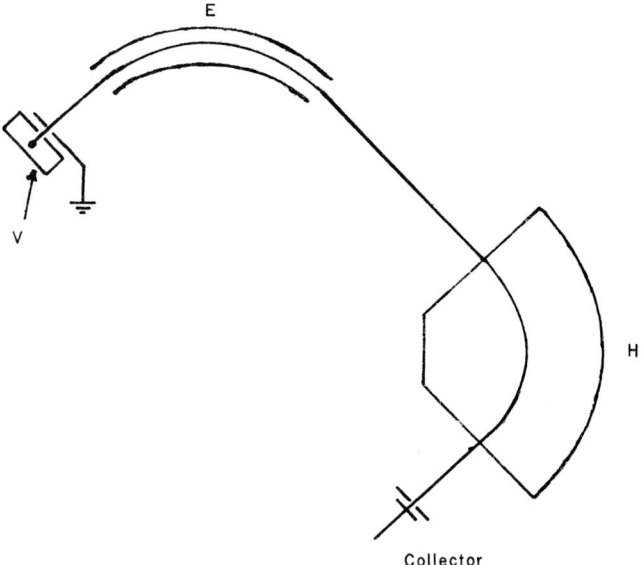

Fig. 1.3. Nier–Johnson double-focusing geometry.

The geometry illustrated in Fig. 1.3 is the Nier–Johnson geometry in which both mass and energy focusing occur at a point. The collector slit is located at this point. The ratio of the electrostatic sector voltage (E) to the accelerating voltage (V) is normally kept at a constant value. However, it follows that, as the double-focusing MS can accept and focus ions of differing energy, the accelerating voltage V which provides the potential energy of the ions, need not be precisely defined, even in its ratio to E, to achieve satisfactory focus. This fact is of importance in voltage peak switching, where the required mass will be brought into focus as soon as the relatively small sector voltage E settles. Thus, it is the much smaller voltage E that must settle rapidly, a point of advantage in peak switching with double-focusing magnetic sector instruments. More recently, double-focusing instruments with reversed Nier–Johnson geometry (i.e. magnet before electrostatic section) have been marketed. The particular uses of these instruments will be described in Section 3.8. Various forms of decoupled operation in which the electrostatic and accelerating voltages are operated independently to facilitate the detection of metastable ions are possible with both double-focusing geometries. The use of these methods will also be described in Section 3.8.

1.2.3 Double-focusing Magnetic Sector Instruments (Mattauch–Herzog Geometry)

The Mattauch–Herzog double-focusing geometry (Fig. 1.4) is unique in that it accomplishes double-focusing for all masses simultaneously. The points of focus lie in a plane so that a focal plane detector such as a photoplate is able to provide an integrated spectrum throughout the mass range. Unfortunately, data acquisition from a photoplate is much more complex and laborious than electrical detection. However, new developments in focal plane detectors (p. 43) promise convenient real time data conversion as well as highest sensitivity.

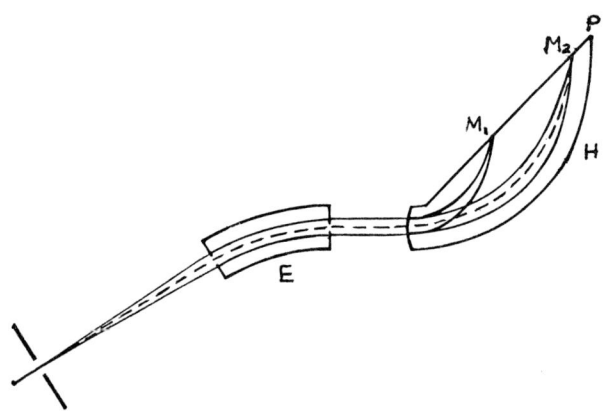

Fig. 1.4. Mattauch–Herzog double-focusing geometry showing simultaneous focusing of different masses (E, electrostatic sector; H, magnet; P, focal plane). (Adapted from ref. 13.)

1.2.4 Quadrupole Mass Spectrometers

The quadrupole mass filter basically consists of four focusing rods arranged as shown in Fig. 1.5. A voltage made up of a d.c. component U and a radio frequency (RF) component $V \cos \omega t$ is applied between adjacent rods. Opposite rods are electrically connected. For particular values of U and V, most ions, having been injected into the filter with a very small accelerating voltage,

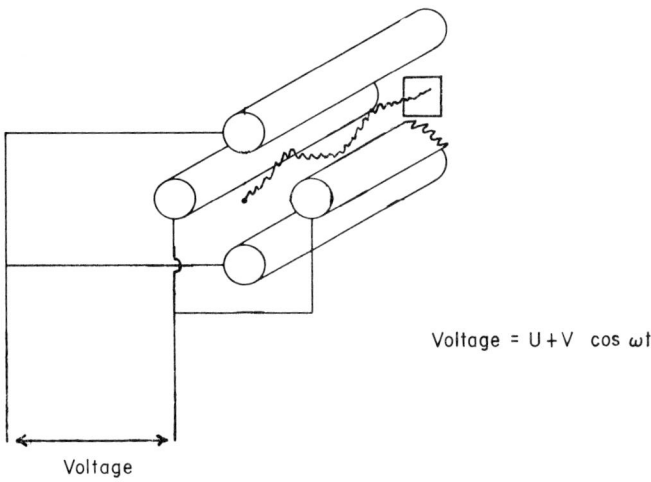

Fig. 1.5. Schematic diagram of a quadrupole mass filter.

undergo an oscillating trajectory of increasing amplitude and are ultimately collected on the rods. Ions of a certain mass, however, can be made to pass through the quadrupole filter. Mass scanning of a quadrupole is accomplished by varying the voltages U and V together, so that the ratio U/V remains a constant. The scan law is then:

$$m = m_o + kV. \tag{1.4}$$

Unlike the magnetic sector MSs, the scan of a quadrupole is highly amenable to computer control. A digital to analogue converter (D/A converter) directed by the computer provides small voltages which are then converted by the quadrupole power supply into specific RF and d.c. voltages corresponding to the transmission of a specific mass.

1.2.5 Time of Flight Mass Spectrometers

Ions produced in the source are drawn into the acceleration region by means of a pulsed extraction voltage and are then accelerated into a field-free drift tube (length, l) by means of an accelerating voltage (V) of a few thousand volts (Fig. 1.6). The velocity attained and hence the time taken to reach the

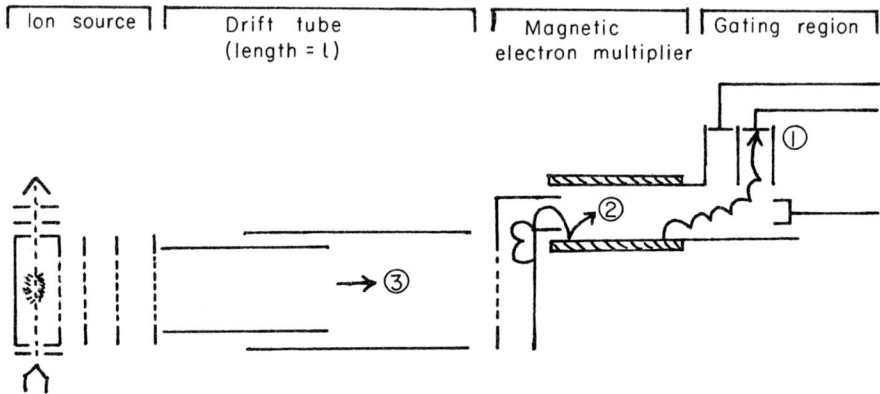

Fig. 1.6. Schematic diagram of a time of flight MS. A lighter ion 1 has been recorded by using the appropriate delay time for gating, ion 2 has just arrived at the collector and the heaviest ion 3 is still in the drift region. (Adapted from ref. 14.)

collector assembly depend upon the ionic mass. Thus, the measurement of drift time effects a measurement of mass (eqn 1.5):

$$t = \left(\frac{ml^2}{2eV}\right)^{1/2}. \tag{1.5}$$

Since the time between adjacent ion bundles is measured in nanoseconds, a time of flight MS with a fast response recording system, e.g. electron multiplier and oscilloscope, is capable of very fast repetition rates (10^4 s^{-1}). For more conventional operation, a gating system is used to direct ions of a specific mass to the collector. By continuously varying the delay time for this gating operation, the mass spectrum can be scanned at rates compatible with more conventional permanent recording systems.

1.3 THE MASS SPECTROMETER COMPUTER INTERFACE

The primary function of any MS computer interface is to accept, condition and digitize analogue data signals from the MS and pass the digitized output to the computer. However, because of the importance of the MS beam data and because of the relatively complex signal conditioning procedures needed for this data prior to digitation, the beam data is usually accorded absolute priority and given its own digitization and analogue to digital (A/D) conversion circuitry. Other analogue signals from the MS, e.g. total ion current (TIC) data, source temperature and electron voltage, are selected for digitization, often by a slower A/D converter, when required by means of a multiplexor. The multiplexor is, in effect, a set of analogue switches controlled by the computer that select the input for digitization on a one at a time basis.

The interface schematic shown in Fig. 1.7 is only one of many possible arrangements. The principles and practice of A/D conversion as well as the transmission and conditioning of the analogue data are considered in Chapter 2 and will not be described here. The other essential function of the interface is as an intermediary in MS control. The digital input/output (I/O) unit produces digital signals to command the MS or other equipment and accepts digital signals from the MS or other equipment. For example, on command from the CPU the digital I/O unit can provide an output capable of energizing a relay in the MS and thus initiate the MS scan. Conversely, the unit can recognize a digital input from the MS signifying that the scan has been terminated, for example, by an end-of-scan mass limit set on the MS.

The D/A converter provides the MS with a variable analogue voltage corresponding to a digital input determined by the computer as a result of program instructions. This facility is used for example to control mass selection by a quadrupole mass filter via a voltage applied to the rods (Section 8.7). Another application is the control of mass selection in magnetic sector MSs via the accelerating voltage (Section 8.6).

1.4 THE COMPUTER

1.4.1 The Central Processing Unit

The interrelationship of the central processing unit (CPU) with other areas of the computer is shown in a very generalized form in Fig. 1.8. The CPU provides calculating ability and controls the overall sequence of operations in the computer. The operation of the CPU is based on the execution of simple instructions (instruction set), that in sequence make up each operation required by a program.

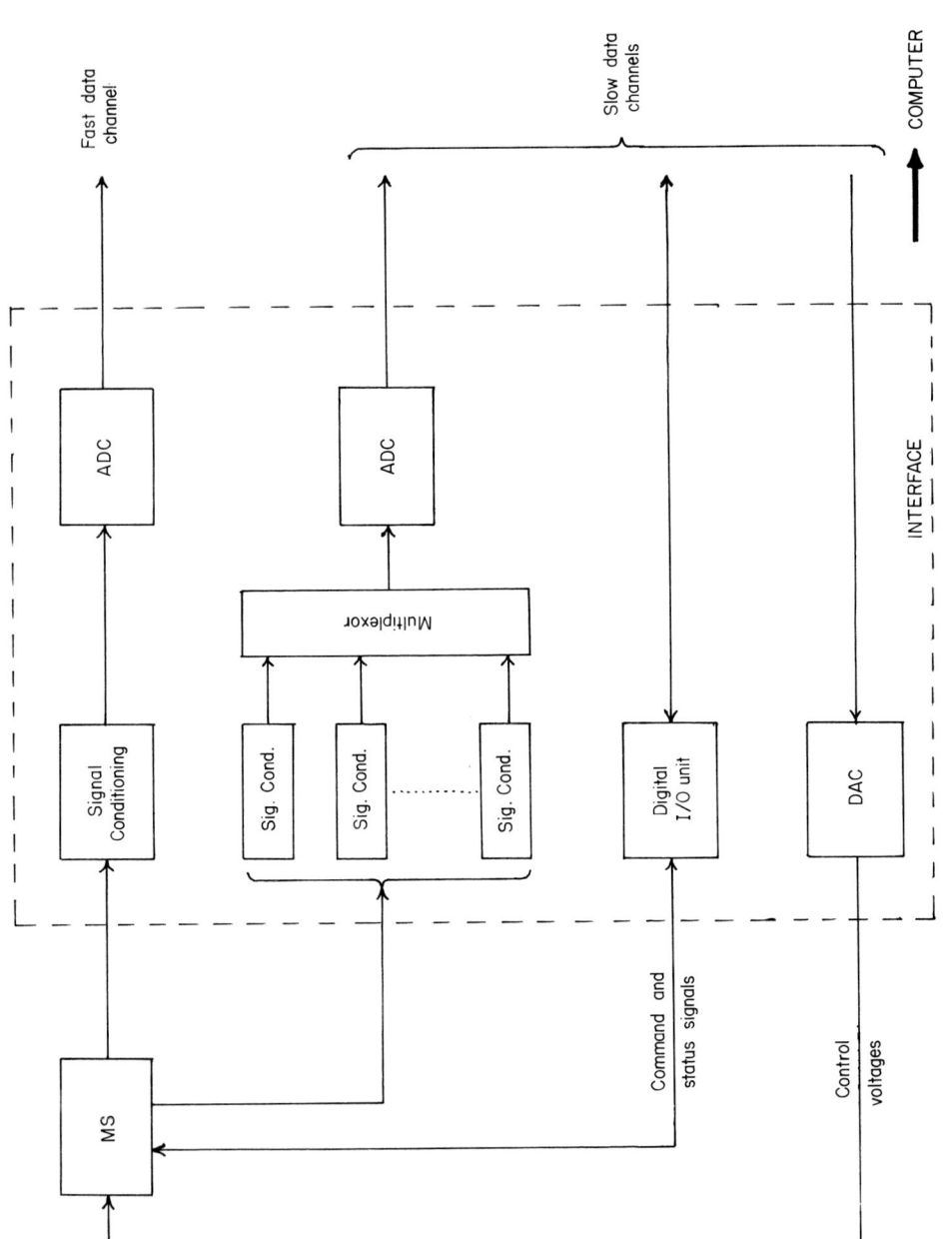

Fig. 1.7. MS computer interface schematic.

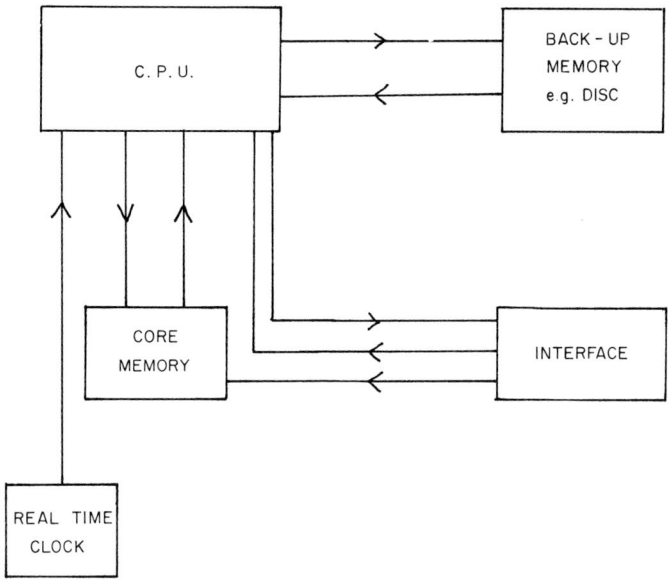

Fig. 1.8. Interrelationship of the CPU with other areas of the computer.

1.4.2 Data Transfer

The interface between core memory and peripheral devices provides conversion and buffering of data undergoing transfer together with control signals to and from these devices. Buffering of data is provided to make the I/O of information compatible with the operational speed of the core memory. Direct transfer from a device as slow as a teletype, for example, is not practical with a memory of this type. Instead, the data characters are stored in a buffer until a word is accumulated, when it is transferred to memory at the normal internal transfer rate of the computer. Control signals are necessary to synchronize operation of the peripheral devices with operation of the computer. For example, in an output routine, the computer must be aware of what information the teletype has already printed out at each moment.

Transfer of data between peripheral devices (including the MS) and core memory may occur indirectly via the CPU under program or interrupt control in a programmed transfer or directly by means of an autonomous data transfer. In data transfer under program control, because the data transfer rate of most peripherals is well below the computer's speed of operation, a great deal of computer time is wasted while continuously checking to see, for example, when one print operation has been completed so that the next one can be initiated. Under interrupt control the computer is made aware, by means of an interrupt, of the fact that a peripheral needs servicing at that exact moment; for example it may be ready to print or it may have finished printing. The interrupt causes the current program to be stopped and servicing of the peripheral is carried out. If there is more than one interrupt in a system, each of these is assigned a priority and servicing is then carried out on this priority basis. In addition to external interrupts from peripherals, interrupts can be written into current programs so that, for example, a cyclic interrupt could cause information to be transmitted at set time intervals to the computer. Although interrupt control is more complex than program control and some sacrifice of data transfer rate is therefore involved, nevertheless it allows much more efficient use of computer time. The computer can be executing background programs between interrupts instead of waiting for data.

In programmed data transfer, the transfer of each set of data requires the execution of several instructions so that the transfer time is at least an order of magnitude greater than the memory cycle time. In autonomous data transfer, the core memory is isolated from the CPU and a direct path from peripheral to core memory established for data transfer. By "stealing" core memory cycles from the CPU, one block of data can be transferred in each core cycle (cycle times are approx. 1 μs in most mini-computers). Autonomous data transfer only requires a sequence of instructions to initiate data transfer; once initiated data transfer proceeds to completion and is not under program control. Alternative names for autonomous data transfer are direct memory access (DMA) and cycle-steal. Obviously DMA is the method of choice for high speed data transfer from a fast scanning MS to a data system. In addition it may be used for data transfer from high-speed peripheral bulk memory devices, e.g. magnetic disc.

1.4.3 Core Memory

Core memory is used to store instructions and data relevant to and provide working space for the program running in the computer at the time. The information is stored in binary units (bits) arranged in blocks or words of n bits where n is the computer word length. Each word has an address associated with it. Core memory is a high speed device which allows rapid data and instruction movement within the computer. Unfortunately, no single memory unit can at present offer both bulk storage and high access speed, so that core memories are relatively low capacity, high speed memories. They are also random access stores. Random access stores permit the retrieval of information in any desired sequence; moreover, the retrieval time is independent of the location of the information in the memory. The speed with which information may be fetched from or stored in memory is the memory cycle time and is usually of the order of 1 μs. It is defined as a read–write cycle time because information fetched from core memory has to be rewritten after use.

The technology of LSI has also made available semi-conductor or solid-state memory. Random access semi-conductor memory (RAM) based on metal oxide semi-conductor (MOS) circuitry is available with the characteristics of fast access and volatility. Volatile memory loses its contents when power is removed. Another form of semi-conductor memory is read only, non-volatile memory (ROM) (see p. 15). This portion of memory is pre-programmed and cannot be changed. The storage capacity of core memory is expressed in units of 1024 (i.e. 2^{10}) words, abbreviated simply to K. For typical performance data see Table 1.1.

1.4.4 Back-up Memory

The back-up memory provides a slower access extension to the core memory for bulk storage of data. The main forms of bulk storage which still retain a moderate access speed and are therefore usable for the on-line storage of mass spectrometric data are magnetic tape and magnetic disc.

Magnetic tape consists of a long strip of flexible base material on which a thin layer of magnetizable material has been deposited. The tape is pulled across fixed magnetic read–write heads by a transport mechanism which is also capable of starting and stopping the tape within a few ms of demand. This is quite different from a disc system which is continuously rotating. Individual words are stored in seven or nine tracks along the tape in a format which depends on the recorder being used. The information is usually segmented into blocks or pages. Access time can be long and depends on where the information is physically located on the tape. Magnetic tape is particularly suitable for the more permanent storage of large quantities of information.

Magnetic discs are thin circular plates coated with magnetizable material. One or more of these discs is mounted on a drive shaft with sufficient spacing between adjacent discs to allow insertion of a magnetic read-write head. Data recorded onto magnetic discs is organized into tracks and these tracks are segmented into blocks or pages. There are two basic types of disc known as fixed head and moving head discs. Fixed head discs usually have one head per track. The disc is not removable from the drive mechanism. Nearly all modern designs use the "flying head" principle; the head mounting permits the heads to ride on a cushion of air when the disc revolves at full speed (typically 2400 rev/min) at a distance of about 0·0025 mm (100 μin) from the surface. The average access time (i.e. the time taken for the head to be in the required position to read or write), is the time for half a revolution (see Table 1.1).

With moving head discs there is usually one head per surface which is placed over the appropriate track by a very precise servo-mechanism. The access time is longer than on fixed head discs because of the time required to position the head. The discs are, however, removable and exchangeable between drives of compatible design, and packs of moving head discs are a common means of providing increased cheap storage capacity. Typical performance data for mini-computer disc, tape and core memory are given in Table 1.1.

TABLE 1.1
Typical performance data for storage devices

	Core memory	Fixed head disc	Moving head disc	Magnetic tape
Capacity in words	4K–128K	$(0 \cdot 25 – 1 \cdot 5)\ 10^6$	$(2 \cdot 0 – 10)\ 10^6$	$(0 \cdot 05 – 20)\ 10^6$
Average access time in μs	0·4	10^4	$(3 \cdot 0 – 7 \cdot 0)\ 10^4$	$10^4 – 10^9$ (depends strongly on location on tape)
Transfer rate in 10^6 bits s^{-1}	20	3·5	2·5	0·04–0·5

1.5 INPUT–OUTPUT DEVICES

Of all the peripheral hardware, input-output (I/O) devices will be most familiar to the user of a data system. These devices are the means by which the user communicates with the data system and is, in turn, communicated with by the system. The MS is also an input device.

A teletype or similar keyboard is used to address the computer and is acceptable for slow output of permanent copy. A visual display unit (VDU) offers by far the most efficient means of user interaction with a data system and greatly increases the flexibility of the system. A third requirement is generally for some faster output device that will also provide graphics, e.g. plotted spectra. Here, a choice may be made from a large range of printer/plotters or a hard copy unit for the VDU. Microfilm photography has also been used to provide permanent records (12).

1.6 SOFTWARE

The term software refers to the programs, i.e. sequence of instuctions, by which the computer is told which tasks to perform. The working language of all computers is referred to as machine language and is composed entirely of numerically coded instructions. A very simple program could be coded by the operator in machine language, the instructions loaded into successive core memory locations by means of the switch register and then the computer told to start executing the program at the starting address. Such an exercise is obviously very tedious, although the ability to use the switches in this way can be very useful, particularly in the investigation of fault conditions. Program writing at the machine language level is usually done in what is termed assembly language. This language is essentially a simple mnemonic representation of machine language that may be typed at the teletype keyboard. Translation into machine code is achieved by a standard program known as the assembler.

It will be appreciated that the operations comprising the instruction set are extremely simple. Thus, a process such as the division of two integers requires the execution of a number of these instructions and may take perhaps 35 µs to complete. Standard operations such as multiplication and division may be greatly speeded by the incorporation of hard-wired logic in the CPU (e.g. hardware multiply divide) requiring only a single instruction for execution which is then completed in 5–6 µs. A newer departure in this field is firmware. In this, rather than hard-wiring a function such as multiplication, it is stored as a routine, usually in a special read only memory (ROM) offering very rapid access times. Most modern mini-computers now offer these facilities.

Since low level or assembly language programming becomes very tedious and requires programmers with considerable experience to handle more complex routines, higher level languages such as Fortran have been developed. In Fortran, the statements used in programming are mathematically and logically familiar and, as such, are far more easily used by non experts. A Fortran program written at the teletype keyboard is translated into machine code (each Fortran statement becomes a block of machine language statements) by a program called the compiler. Extensive standard editing programs allow the simple correction of Fortran programs during their development.

However, higher level languages also have their disadvantages. Programs compiled from high level programs are less efficient in their use of computer memory and computer time. Thus, high level languages are not appropriate for on-line data acquisition and, in general, they are not appropriate when time and memory are at a premium. However, the complexity of assembly language programming must always be balanced against these considerations. A further possible disadvantage of high level programming, is that the compiler itself occupies a considerable amount of core memory in use.

Another type of high level language is available for use. This is the interpreter-based language such as Basic or Focal. Unlike a compiler, the interpreter translates and then immediately executes the program line by line. The language and routines available are very similar to those available in Fortran (see Fig. 1.9). However, the combined translation and execution

```
01.01 E
01.05 F I=1,1,4;T !
01.20 S I=0
01.21 T !;S I=I+1
01.22 D 10
01.25 I (I-3)1.21,1.21,3.10

03.10 S A=T(I-3);S B=T(I-2);S C=T(I-1)
03.20 S D=(A-B)/FLOG(M(I-2)/M(I-3))
03.30 S E=(B-C)/FLOG(M(I-1)/M(I-2))
03.40 S G=E-(D-E)*((M(I-2)-M(I))/(M(I-3)-M(I-1)))
03.50 S H=C-G*FLOG(M(I)/M(I-1))
03.60 T %8,H
03.62 S P=T(I)-H
03.68 T "        ";T %4,P*1E6/G
03.69 T "        ";T %6,G/100
03.70 G 1.21

10.05 T "M(";T %2,I;T ")   "
10.10 A M(I)
10.15 T "      T(";T %2,I;T ")   "
10.20 A T(I)
*
```

Fig. 1.9. Focal program for the computations of eqn. 3.6.

make the system completely interactive with the operator. Thus, this system is probably the simplest of all to use in the editing and development of high level programs. However, as the interpreter is in use at all times, execution is relatively slow and, because the interpreter is in core at all times, the size of the program and number of variables that can be accommodated is restricted.

REFERENCES

1. McMurray W. J., Green B. N. and Lipsky S. R. (1966). *Analyt. Chem.* **38**, 1194.
2. (a) Olsen R. W. and Burlingame A. L. (1965). 13th Annual Conference on Mass Spectrometry, St. Louis. Paper 37, p. 192.
   (b) Hites R. A. and Biemann K. (1967). *Analyt. Chem.* **39**, 965.
3. (a) Biemann K., Bommer P., Desiderio D. M. and McMurray W. J. (1966). *In* "Advances in Mass Spectrometry" (Ed. Mead W. L.), Vol. 3, 639. Institute of Petroleum, London.
   (b) Burlingame A. L. (1966). *In* "Advances in Mass Spectrometry" (Ed. Mead W. L.), Vol. 3, 701. Institute of Petroleum, London.
4. Ziegler E., Henneberg D. and Schomberg G. (1970). *Analyt. Chem.* **42**, 51A.
5. Klopfenstein C. E. (1974). *J. agr. Fd Chem.* **22**, 736.
6. Baumann F., Hendrickson J. and Wallace D. (1974). *Chromatographia*, **7**, 530.
7. Dessy R. E., Van Vuuren P. J. and Titus J. A. (1974). *Analyt. Chem.* **46**, 917A.
8. Klein P. D., Haumann J. R. and Hachey D. L. (1975). *Clin. Chem.* **21**, 1253.
9. Mass Spectral Search System, Cyphernetics International Corporation.
10. McLafferty F. W., Dayringer H. E. and Venkataraghavan R. (Unpublished).
11. Carhart R. E., Johnson S. M., Smith D. H., Buchanan B. G., Dromey R. G. and Lederberg J. (August, 1975). 170th Annual Meeting of the American Chemical Society, Chicago.
12. (a) Biemann K. (1972). *In* "Biochemical Applications of Mass Spectrometry" (Ed. Waller G. R.) p. 96. John Wiley and Sons, Chichester.
    (b) Biller J. E., Hertz H. S. and Biemann K. (1971). 19th Annual Conference on Mass Spectrometry and Allied Topics, Atlanta. Paper F2, p. 85.
13. Mattauch J. and Herzog, R. (1934). *Z. Phys.* **89**, 786.
14. McFadden W. H. (1973). "Techniques of Combined Gas Chromatography-Mass Spectrometry: Applications in Organic Analysis". John Wiley and Sons, Chichester.

2
Data Acquisition

2.1 INTRODUCTION

Following the description of the mass spectrometer (MS) and the components of the data system we can return to the system as a whole and consider the treatment of data originating from the MS. The basic processes around which this chapter is written are firstly, the conversion of the analogue output of the MS to a computer compatible, digital form and secondly, the conversion of this digital data into time and intensity data for each peak. It is convenient at this stage to give separate consideration to data acquisition from scanning MSs and from focal plane MSs. (see Section 2.5.)

2.2 DATA ACQUISITION FROM SCANNING MASS SPECTROMETERS

2.2.1 General Methods

There are at least four potentially feasible methods for acquiring digital data from the analogue output of a scanning MS. These may be summarized as follows:

(i) Dedicated on-line computer.
(ii) Time-shared on-line computer.
(iii) Frequency modulated (FM) analogue magnetic tape and remote computer.
(iv) On-line digitizer, digital tape and remote computer.

Potentially the most attractive and certainly the most common system is the use of an on-line mini-computer dedicated to the needs of one, or perhaps more, MSs. The continued improvement in mini-computer specifications are more than adequate to meet the great majority of requirements in mass spectrometry. Modern commercial mini-computer based systems are able to acquire data and convert it to mass and intensity information on-line at even the fastest scanning speeds demanded by gas chromatography mass spectrometry (GCMS) operation. Graphical or tabular summaries of the scan data are presented during the experiment in real time and enable the operator to both monitor the progress of an experiment continuously and interact with the system when required. An increasing trend towards computer control of MS operation, particularly in the quadrupole field, is also best served by the use of a dedicated on-line computer. As a result of its popularity, the dedicated on-line system is used in the subsequent pages as a model system for a detailed description of data acquisition from a scanning MS. Again, in this description, the use of a continuously sampling analogue to digital converter (A/D converter) is assumed as the most common means of producing digital data. Discontinuous methods of sampling with an A/D converter are discussed on p. 36 and an alternative method, using ion counting, is discussed in Chapter 8.

The use of a time-shared computer on-line has developed in some larger installations where central computing facilities are available to service an extensive range of analytical instrumentation. There are some advantages in the use of a central computing system, such as the access to very powerful computing facilities and the reduction in effective cost of expensive peripheral devices when these are shared by a number of analytical instruments. However, a modern MS produces data at a far greater rate than most other analytical instruments in common use. This means that, certainly for the whole of the scanning period, the MS has to have absolute priority over other instruments serviced by the system. This demand imposes a considerable strain on the equitable operation of such a system. For this reason, a MS is sometimes interfaced to a smaller dedicated satellite computer that acquires data to transfer to the central computer. This is the basis of the so-called hierarchical system (1).

An alternative means of using the facilities of a central time-sharing computer system is to initially record the mass spectral output onto FM analogue magnetic tape and this data tape is then hand-carried to the computer for digitization and further off-line processing (2). Three channels of data may be recorded on separate tape channels at different attenuation factors, e.g. × 1, × 10, × 100, to overcome the limited dynamic range of the tape. When playing back the data into the computer, the least attenuated channel that is not saturated, is selected by hardware for digitization. Further channels can be used for scan synchronization and to record timing pulses alongside the mass spectral data to minimize errors which are due to tape stretch and speed variation on replay. The use of timing pulses in this way is necessary when attempting accurate mass measurement of data recorded on analogue tape.

Use of a central computing facility in this manner permits efficient use of computer time as digitization, peak time calculation, time to mass conversion and any other data reduction processes can be carried out in a batch mode in off-peak hours more or less without intervention by the operator. Another historical reason for the use of analogue tape-recording techniques was to provide the high digital sampling rates which were necessary, particularly for high resolution and fast scanning operations, since these were not directly available to early workers in this field (3). This was achieved by recording data at a high tape speed (e.g. 60 inches per second (in/s)) and replaying at a slower speed (e.g. 3·75 in/s). Thus, an actual digitization rate of 2 kHz on replay gave an effective digitization rate of 32 kHz. Another advantage of a complete record of high resolution scans is that it makes possible further processing such as deconvolution of unresolved multiplets should it be necessary.

The alternative tape-recording method involves initial digitization followed by recording on digital magnetic tape (4). Again, a complete record of the scan data is produced. However, this method leads to shorter tape records than the use of analogue tape particularly when a digitizing device, such as the Carrick interface which produces only one intensity reading and one time reading per peak, is used. Some idea of the length of magnetic tape required to record a set number of scans using different tape-recording methods may be gained from Table 2.1. For GCMS operation where long runs containing more than 1000 spectra may need to be recorded, the economy of storage offered by Method 3 (see Table 2.1) makes it the most attractive system if tape-recording is required.

TABLE 2.1

Comparison of efficiency of various digital tape-recording methods for time and intensity data

Data acquisition method	Length of tape/100 scans (ft)
(1) A/D conversion written directly to magnetic tape	800 (244 m)
(2) Thresholded A/D conversion—magnetic tape	150 (46 m)
(3) Carrick Interface—magnetic tape	15 (4·6 m)

Time word is not written to tape in Method 1 as it is implicit in the serial order of intensity words. Conditions: 300 peaks/scan; scan speed, 3 s/decade; mass range, m/e 12–600; resolving power 1000; tape density, 800 b.p.i.; A/D conversion rate, 12 kHz. (Data taken from ref. 4a.)

In any comparison of the four data acquisition methods described, the overwhelming advantage in favour of the dedicated on-line computer system, particularly compared with the off-line systems, has been the immediate availability of results. Besides the rapid return of processed analytical data, the user of an on-line system is also made immediately aware of most faults (e.g. decomposition of the sample, development of an instrument fault) so that when remedial action is possible, the time wasted in a useless analysis is cut to a minimum.

2.2.2 Electron Multiplier and Pre-amplifier

On introduction of a sample into the source of a MS, ions characteristic of the sample are produced. These are extracted from the source region and separated according to their mass to charge (m/e) ratio. At different masses, ion currents of widely differing intensity will be sampled at the collector region. These may range from more than 10^{-10} A down to minimum values determined by the application, perhaps about 10^{-18} A in organic mass spectrometry (see p. 236). The most widely used form of detection and amplification of these currents is the electron multiplier. Most multipliers comprise some 12–20 dynodes electrically connected through a resistive network (see Fig. 2.1). The ion beam strikes the first dynode which acts as an ion to electron converter. The secondary electrons produced are then accelerated towards the second dynode where their impact causes an amplified electron emission and this process is repeated for each stage of the multiplier. The gain of the electron multiplier, which depends on the applied multiplier voltage, is produced with virtually no noise and with negligible time constant. The final dynode of the electron multiplier is connected to a pre-amplifier (see Fig. 2.1) which converts the output current into a voltage suitable for digitization or for recording on a fast recording device, such as an ultraviolet recorder.

A pre-amplifier circuit incorporating an operational amplifier (5) is shown in Fig. 2.1. In this circuit, the resistor R_f provides a feedback path between the amplifier output and input. As the output voltage is of the opposite sign to the input, feedback through R_f tends to decrease the input signal to the amplifier. In fact, because of the high gain of the amplifier, the input signal tends towards zero and the amplifier input becomes a virtual earth point. A d.c. offset voltage to compensate for drift is provided at the non-inverting input.

The output voltage of this amplifier is the product of the multiplier current and the resistance R_f. For example, in the circuit illustrated, an input of 10^{-8} A from the multiplier final dynode results in an output of 0·5 V. The maximum output voltage of most operational amplifiers is ± 10 V. The time constant of the circuit τ is the product of the feedback resistance R_f and the stray capacitance C_f across this resistor. With R_f equal to $5 \times 10^7 \, \Omega$, the value of τ is approximately 5×10^{-6} s, so that the amplifier can deal with signals containing frequencies up to 30 kHz ($f = 1/(2\pi\tau)$).

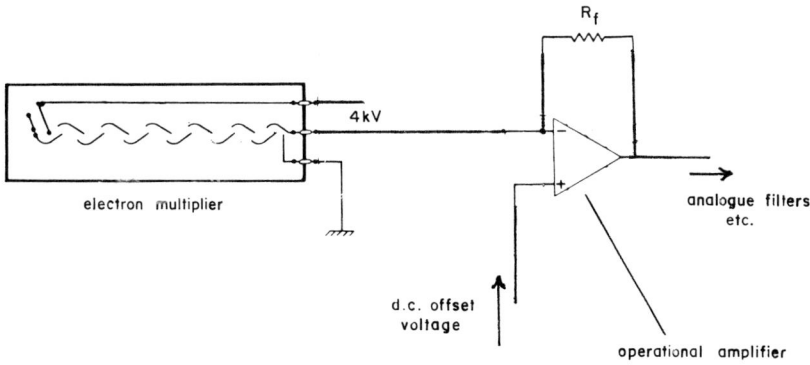

Fig. 2.1. Electron multiplier and pre-amplifier circuit.

2.2.3 Analogue Signal Conditioning

The output of the MS, taken from the pre-amplifier, will invariably have noise originating in the amplifier impressed upon it. In addition, the statistical nature of weak ion currents means that peaks recorded under these conditions may have very irregular profiles. A suitable analogue filter, designed to pass only signals with a frequency less than a specified cut-off frequency, will reduce this noise level and will smooth the peak profiles. Thus, an A/D converter, only able to take a limited number of samples of the voltage output, will be able to provide a better digital representation of the peak profile after adequate smoothing.

It has been shown by Banner (6) that the amplifier time constant τ should lie between $t_p/10$ and $t_p/20$, where t_p is the peak width, in order to keep the loss in resolution and peak height on scanning at a figure between 5% and 10%. As an example, for an exponential scan, $t_p = t_{10}/2 \cdot 303R$, giving a peak width for a 1 s scan at 1000 resolving power of approx. 4×10^{-4} s, so that τ in this case should lie between 2×10^{-5} and 4×10^{-5} s. Thus, the filter used should have a cut-off frequency of approximately 5 kHz ($f = 1/2\pi\tau$).

2.2.4 Signal Transmission

In much existing instrumentation, the A/D converter is located within the computer, usually distant from the MS. Thus, the amplified-filtered analogue output must be transmitted to the computer location. A suitable transmission technique involves the use of shielded cable for increased noise immunity. However, the transmission of analogue signals over any distance is always attended by the problem of noise pickup in the transmission lines and this can be further aggravated if the signal earth is different from the computer earth, introducing an earth loop.

An alternative method, which is becoming increasingly common, is to digitize the data at the MS site and then transmit the data to the computer in digital form. Such a system offers the advantages of the high noise immunity of digital data transmission and an A/D converter dedicated to the instrument.

2.2.5 Signal Multiplexing

After the signal conditioning stage, an analogue multiplexor may be inserted. The multiplexor is a solid-state device, generally operated under computer program control which selects the analogue inputs on a "one at a time" basis for onward transmission, in this case to the A/D converter. The multiplexor has a number of possible functions. It may be used to select the appropriate channel where a single MS beam input has been divided into two or more channels, each employing a different gain, in order to extend the effective dynamic range of the A/D converter (cf. p. 237). A multiplexor is also used when more than one independent source of analogue data has to be sampled by the same A/D converter. For example, the total ion current (TIC) signal from the MS may be multiplexed with the beam data and then sampled under program control using the same fast A/D converter. Other analogue signals, e.g. a Hall probe output, can be multiplexed in the same way. Alternatively, these subsidiary analogue signals can be multiplexed and then sampled using their own, slower A/D converter.

2.2.6 Analogue to Digital Conversion

Using a time base derived from a crystal oscillator and acting under computer control the A/D converter performs digital conversions of the input analogue voltage at precisely defined time intervals (see Fig. 2.2). The computer thus receives a series of digital signals representing the incoming analogue signal.

The most common type of fast A/D converter is the successive approximation converter. In this the analogue input is compared to a series of precise analogue voltages produced by a D/A converter. Its operation is most easily understood by reference to Fig. 2.3 in which are shown all the possible number combinations generated by a successive approximation A/D converter with 4 bit resolution.

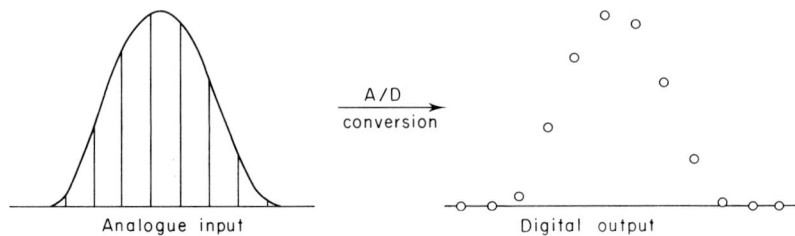

Fig. 2.2. Operation of an A/D conversion.

A typical conversion is underlined for an analogue input that is equal to a digital value between decimal 10 and 11 (i.e. binary 1010 and 1011). At the beginning of the conversion the analogue input is compared with binary 1000. Since the input is greater than that, the logic now effects the comparison with a voltage in which the second binary digit has become a "1", i.e. 1100. Since the input is now smaller than this voltage, the second binary digit is return to "0", the third digit becomes "1" and a comparison is again effected with 1010. This is smaller than the input, so, the fourth digit becomes "1" and comparison is made with 1011. Since this is greater than the input the final step returns the last digit to "0", and the digital value 1010 is fed into the computer as the nearest equivalent to the analogue input. For an n bit successive approximation converter there are always n comparison steps and therefore the conversion time is fixed, irrespective of the input voltage, an important consideration in the rapid A/D conversion of a MS output.

In considering the specification for an A/D converter, some idea of the minimum size of signal to be digitized may be gained from the fact that the smallest recognizable peaks can present a profile that, for a considerable portion of its width, has a height no greater than that of a single ion pulse (cf. p. 238). For a 1 s/decade, 1000 resolving power exponential scan, recorded using a filter with a cut-off frequency of 5 kHz (cf. p. 22), a single ion pulse typically has a base width of $1 \cdot 2 \times 10^{-4}$ s and therefore the height will be equivalent to $(1 \cdot 6 \times 10^{-19} \times 2)/(1 \cdot 2 \times 10^{-4}) = 2 \cdot 7 \times 10^{-15}$ A at the collector. With an amplifier input resistor of $5 \times 10^7 \Omega$ (cf. p. 21) and a multiplier gain of 10^6, a single ion pulse would have a height of 133 mV (this

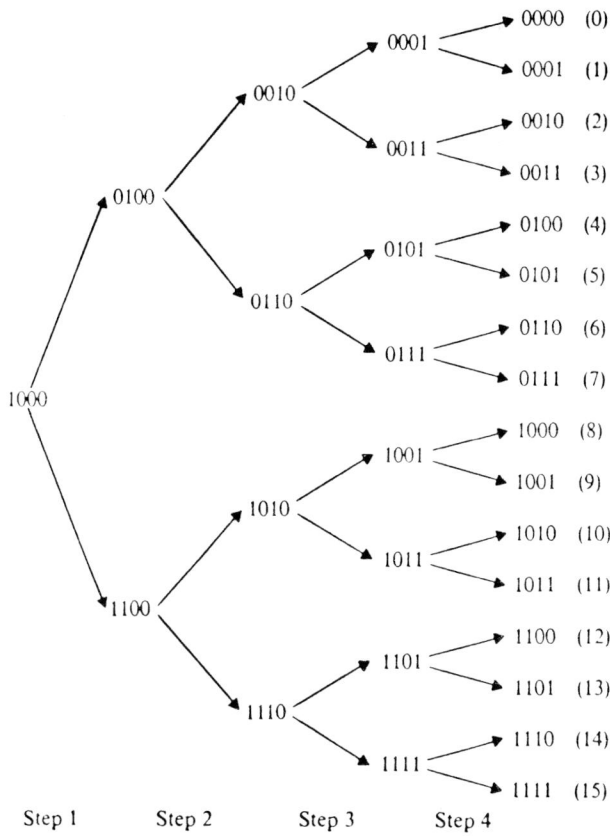

Fig. 2.3. Possible pattern generation for 4 bit successive-approximation A/D conversion. (Reprinted with permission from ref. 5b. Copyright by McGraw-Hill Book Co.)

is an average figure because of statistical variations in multiplier gain). With a more routinely available gain of 2×10^5, the average height would be 27 mV.

If the scanning conditions are relaxed to 10 s/decade, 1000 resolving power, then the lower bandwidth used will result in a single ion pulse with an average height of 2·7 mV at a gain of 2×10^5. The amplifier noise level may well approach 1 mV peak to peak under these conditions, so that the use of a higher gain would afford a better signal to noise S/N ratio. However, with a maximum output of 10 V, the use of a higher gain also decreases the dynamic range of signal that can be recorded. The foregoing calculations illustrate the need, under scanning conditions, for an A/D converter with a resolution of the order of 1 mV or better. Thus, for a maximum signal amplitude of 10 V, the commonly used 12 bit and 14 bit fast A/D converters give resolutions of 2·5 mV and a more satisfactory 0·625 mV respectively. As will be seen in Chapter 8, relaxing the recording conditions still further to include single and multiple peak monitoring (MPM), results in a requirement for a still greater dynamic range.

An indication of the dynamic range of signals encountered may also be obtained from a consideration of sample sizes and the relative intensity of useful mass spectral peaks. A GCMS analysis in which a 1% component of a mixture is to be identified would not be unusual, nor would the use of mass spectral peaks down to 1% of the base peak. Thus, the data already covers a dynamic range of 10^4 or 15 bits. Some latitude in the magnitude of mass spectral peaks considered to be useful is, of course, available, but the importance of a system that can record a large dynamic range of signal and of software that can recalculate the intensity of saturated peaks (cf. p. 90) is illustrated.

Since the A/D converter has a finite conversion time, Δt, there will be some uncertainty in the time assigned to the digitized value. This value will represent the analogue value at some time between t_0 and $t_0 + \Delta t$ rather than that at t_0. Here, t_0 is the time at which conversion begins and is the time assigned to the digitized value (see Fig. 2.4).

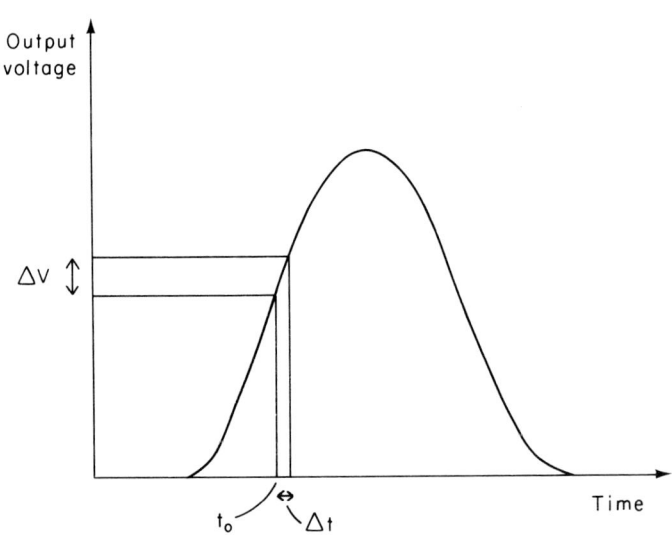

Fig. 2.4. Uncertainty in A/D converter time.

Ideally, the change in the analogue signal during the conversion time should be less than the least significant bit of the A/D converter. As the conversion time of a fast 12 bit A/D converter is several µs (say, at least 5), a 10 V signal recorded at a resolving power of 10000 and a scan speed of 10 s would change in this time by at least 250 mV, far more than the least significant bit. The effect of these misdeterminations of the analogue voltages is to shift the whole peak in time and therefore in apparent mass (see Fig. 2.5). The maximum shift that could be expected is about 1 p.p.m. for a 10 s, 10000 resolving power scan and about 10 p.p.m. for a 1 s, 1000 resolving power scan. To alleviate this problem, the A/D converter is usually preceded by a sample and hold amplifier (S/H). This amplifier follows the analogue

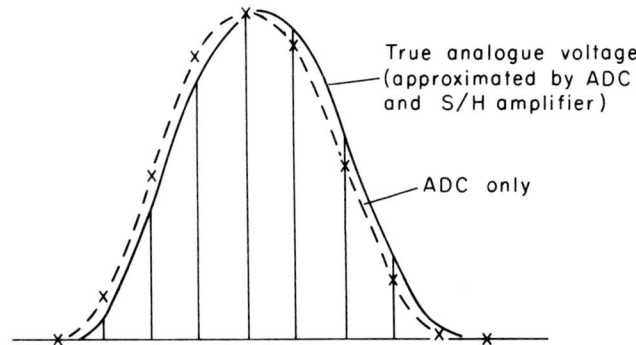

Fig. 2.5. Effect of uncertainty in A/D converter on apparent peak position.

input voltage by allowing the voltage to be stored on a condenser. At the time of digitization the stored voltage is "held" by isolating the condenser from the input by means of a switch. The uncertainty in timing now is only the aperture time of the amplifier, i.e. the time taken for the switch to open. This is usually of the order of 10–100 ns.

The setting time of a S/H amplifier is typically 1 µs to attain within 0·1% of an input change. This time must be added to the A/D converter conversion time which is now a minimum of 6 µs, giving a maximum possible sampling rate of approx. 170 kHz.

The maximum sampling rate that can be used is controlled not only by the specifications of the S/H amplifier and the A/D converter but also by the rate at which digitized data can be further transferred to whatever form of storage is used for data on-line. The digital data may, of course, have been further processed in the course of being transferred to this storage area. For example, when a demand for direct memory access (DMA) is made, some time must elapse until a permissible break point is reached in the program. The transfer itself then takes at least the core memory cycle time. The sampling rate possible is determined by the sum of these times, together with the time needed to compare the digital value with the digital threshold, perhaps a total of 10 μs, or a digitization rate of approx. 100 kHz. However, by interposing a small buffer register, the sampling rate may be increased as the previous sample may be cleared to the buffer register to await DMA and a new conversion may immediately succeed this one. The use of a buffer may also be necessary to retain sufficient digitized data to allow the operation of a peak recognition routine.

Bearing in mind these upper rates fixed by hardware limitations, a choice of a digitization rate suitable to the analytical problem at hand must then be made. In mass measurement, where the mass is derived from an accurate estimate of the peak position in time, usually as its centroid, 10–20 digital measurements over the peak are commonly accepted as necessary to ensure an accurate mass determination. For conventional mass measurement conditions, viz. an exponential magnetic scan of 10 s/decade at 10 000 resolving power, this requires a minimum sampling rate of 25 kHz. On reducing the resolving power, the peak width and sensitivity increase. Thus, the amount of digital data arriving at the computer can increase considerably, particularly if the digitization rate is not reduced. If this is not done, the processing of the data could become rate-limiting. For example, where peak time calculation or time to mass conversion is carried out on-line, there is a limit to the speed with which this processing may be accomplished. Data may be buffered in core pending processing, but if a high intake of digital samples persists for any length of time, the buffer may become full and further intake of samples will not be possible.

Little information regarding the optimum sampling rates for nominal mass operation is available. In the absence of experimental data, it seems reasonable to use approximately the same number of measurements over a peak width (i.e. 10–20) as in the accurate mass case. A rule-of-thumb criterion suggested by Perone and Jones (5b) is that for faithful reproduction of the signal, the sampling frequency should be at least $10 \times$ the bandwidth of the waveform. On this basis, a time constant equal to 1/15 of a peak width (see p. 22) implies a minimum of approx. 20 samples over a peak width. Some 10–20 samples covering a peak width would also appear to be suitable for the subsequent use of a digital smoothing routine (12, 13). As an alternative to selecting a different sampling rate for each set of scanning conditions, the use of a fixed fast sampling rate and the averaging of a pre-selected number of samples to give each data point has been suggested (7).

2.2.7 Interface Logic and Control Functions

The control functions operated by the computer during data acquisition are a vital part of this process. These functions include control of data channel switching, sampling of incoming signals, digital conversion of signals and data transfer in the correct time sequence, counting and recording the number of samples taken during a scan and marking the end of a peak, organizing the sequence in which data is recorded in the computer, monitoring and controlling the scanning of the MS, and recognizing and recording fault conditions.

The timing source or clock, which is at the heart of the system is a crystal-controlled oscillator. The crystal oscillator is the most precise mechanical oscillator and frequency generator available, and errors introduced by the clock into sampling are small even for accurate mass measurement. The A/D converter sampling rate may be set manually or via the software. Essentially the setting determines the number of clock pulses between successive samples. Usually a counter is incremented by one at each clock pulse and when it reaches a value appropriate to the desired sampling rate it returns to zero whereupon an A/D converter cycle is initiated and a counter recording the number of samples is incremented by one. The contents of this latter counter are injected into the data stream to the computer at the end of the peak, usually when the first sample falls below the comparator threshold to establish the position of the peak in the scan. The contents of this counter may be returned to zero or maintained at end of peak depending on the size of the count there. A count can also be made of the number of noise peaks detected, i.e. those "peaks" rejected by one of the tests for peak validity.

Requests for the reading of data by the computer are carried out through interrupts. By means of these signals, the current program operating in the central processor is interrupted, the contents of working registers stored and a jump made to a data reading routine. After the data has been read, a return to the original program is made. Interrupts may originate from a number of sources, e.g. beam data, clock counter data, other data from an A/D converter such as TIC data and scan state signals from the MS. Each interrupt has a priority status so that, for example, beam data is always read with the highest priority.

Recognition of MS states and control of MS operation is effected through the interface by means of digital signals. For example, before initiating clock operation and A/D conversion, the computer must be aware that the MS has commenced scanning. In this case, either a voltage level or a contact closure originating from the MS may be sufficient depending on the design of the interface. Control of the MS may be a simple operation such as initiation of the scan from a teletype keyboard. Alternatively, full control of the scan may be effected as in the use of digital voltages developed by a D/A converter to control the mass selected by a quadrupole mass filter (cf. p. 235).

Control of other parameters such as sample temperature during an analysis is possible as is the automatic checking of resolution and sensitivity before an analysis. Conversely, the MS may act as a control element via the computer. For example, operating routines for processing plant may be determined on the basis of mass spectrometric analyses (8). Recognition of fault conditions pertinent to the operation of the interface is another necessary function. Once recognized, the existence of the fault condition is signified by an interrupt to the CPU so that the appropriate action, even if this only involves the printing of an error message, can be implemented.

2.2.8 Digital Thresholding

A number of data acquisition systems operating on the basis of continuous A/D conversion, particularly those used with high resolution MSs, incorporate the functions of a digital device known as a comparator. The comparator permits each value developed by the ADC to be compared with a set digital threshold and only those values exceeding this threshold are transmitted on to the CPU. The primary function of the comparator is to prevent the transmission of baseline information so as to reduce the amount of digital data that has to be processed in peak recognition routines etc. This is particularly important in high resolution operation when, at a resolving power of 10000 for example, useful data (peaks) probably occupies only about 1% of the total scan time.

The setting of a fixed digital threshold has an important bearing on the electron multiplier gain that must be used to allow detection of the smallest peaks. At very low sample levels, as previously mentioned (p. 25), the ions constituting the scanned mass spectral peak can arrive at intervals sufficiently well separated for the recording system to present the peak as an envelope with a height little different from that of a single ion peak. To ensure detection of such small peaks, the multiplier gain must be sufficiently high for the single ion peak height to be (a) well above the amplifier noise level and (b) above the comparator threshold. If a comparator threshold is employed, the digital samples rejected must also be counted if the time of any recorded peak is to be determined from either that of the preceding peak or the start of the scan.

2.2.9 Peak Recognition Routines and Data Enhancement

The train of digital data above the comparator level results from normal peaks, single ion peaks, noise peaks and metastable peaks. The observed width of a single ion peak is determined by the bandwidth of the recording system but is much less than that of true peaks which we consider as containing at least four ions. Noise peaks are of variable intensity but are similarly narrow compared with normal peaks, particularly in proportion to their height. Metastable ions are ions which have been formed by decomposition of another ion in flight after acceleration. Metastable ions will be discussed in more detail in Section 3.8, but we may say at the moment that they are seen as relatively small, wide peaks which are usually centered at a non-integral mass (Fig. 2.6). Metastable ion peaks and multiplets from unresolved normal ions may both be distinguished from resolved normal ion peaks by their excessive width, particularly in proportion to their height. The main function of any peak recognition routine is to reject noise and single ion peaks. In addition the routine may be used to differentiate between normal and metastable peaks or multiplets (see Chapter 3).

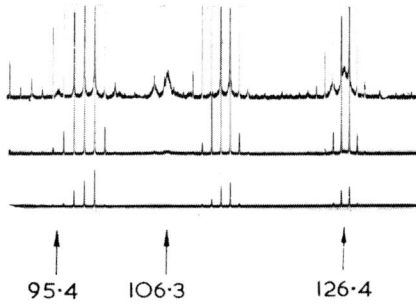

Fig. 2.6. Metastables in the C_7–C_9 region of n-hexadecane.

Some comparators also contain the logic necessary to discriminate between true peaks and noise. In these cases, the logic is necessarily simple and is usually of the kind that requires a small number (about three) of consecutive digital samples to exceed the threshold for a true peak to be recognized and the data processed as such. In other cases, all data above whatever threshold has been set, whether representing a peak or not, is transferred to core. Peak recognition routines carried out in core vary considerably in detail, nevertheless they almost all incorporate two basic tests. These tests are that (a) more than a minimum number of consecutive samples should exceed a specified threshold for the beginning of a peak to be recognized and that (b) more than a minimum number of consecutive samples below the threshold constitutes the end of a peak. In addition, a number of these peak recognition methods employ a preliminary digital filtering or smoothing routine.

Use of a test that requires more than a specified number of consecutive samples to be above threshold for the beginning of a peak may cause some small peaks to be incorrectly rejected (see Fig. 2.7).

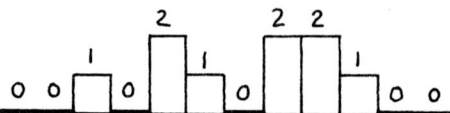

Fig. 2.7. This peak would be rejected if the peak recognition routine required four consecutive samples above the threshold. $0 \equiv$ below threshold.

To avoid this situation, it has been suggested that one or perhaps two consecutive below threshold samples may be allowed to interrupt a train of above threshold samples without causing rejection as a peak. Such a procedure might well be advantageous in offering more certain detection of small peaks although the performance of a system detecting resolved multiplets with the threshold could be compromised in this case.

Perhaps a better method of improving the detectability of small peaks is to use a digital smoothing routine prior to thresholding (7, 9, 10, 11). These computational methods can be designed to improve the S/N ratio by reducing noise that is of a shorter period than the mass spectral peaks themselves. Compared with hardware filtering, they are time-consuming but can be designed precisely to suit the experimental data. In this way, statistical fluctuations which might cause the profile of small peaks to drop below a threshold will be removed to some extent and allow the detection of small peaks by a simple thresholding procedure. The smoothing routine adopted by Hites and Biemann is illustrative (10). As soon as five data points have been collected, they are smoothed using eqn 2.1 where Y_j and Y_j^+ are the original and smoothed data points respectively:

$$Y_j^+ = \tfrac{1}{16} \sum_{i=-2}^{i=+2} C_i Y_{j+i} \, [C_{\pm 2} = 1, C_{\pm 1} = 4, C_0 = 6]. \tag{2.1}$$

Each successive data point Y_j^+ is smoothed using the two preceding and succeeding data points in this way. The more commonly used coefficients described by Savitsky and Golay (9) were found to be less useful. To take advantage of the improved S/N ratio, thresholding should follow smoothing. The use of a smoothing range that does not exceed half the peak width is preferred so as to retain as much of the structure of the original data as possible (12, 13).

In cases where, at the end of a peak, the beam current may not drop below the threshold, particularly with instruments of lower resolving power, the beginning of a peak may be detected by the occurrence of a greater than specified intensity rise from the previous data point rather than a rise above a set threshold (14, 15, 10). In the implementation of this type of system by Holmes et al. (14), the rise was made variable to allow for the increased noise level found with more intense signals. Thus, the rise was set at a fixed value plus one-quarter of the momentary signal value. Most peak recognition routines employ the second basic test, viz. more than a minimum number of consecutive samples below the threshold constitutes the end of the peak, although generally, the end of a peak is signified when only one sample falls below the threshold. Some methods (16) also allow different threshold levels to be set for the beginning and end of the peak.

In cases where a threshold is not employed, the test for the end of a peak can be the occurrence of a pair of equal or increasing samples following the peak maximum (10). Other routines, where the peak position is defined as the position of maximum intensity of the peak and particularly where the MS may be of a lower resolving power, require only that a certain drop in intensity from this maximum occurs for the peak to be recognized and no actual end of peak is detected as such (14, 15). Such routines that measure peak position as the position of maximum intensity or use peak height rather than area as a measure of intensity must also test for and record the sample of maximum intensity. It is also possible to reject noise peaks on a width basis, without using a threshold if the width of a suspected peak profile is calculated at a standard height, say half peak height (15).

An alternative approach to the problem of locating spectral peaks in noise contamination is the use of a cross correlation technique. This method is used to locate a peak shape that is similar to a reference shape represented by a series of intensity values. Such a technique for real time operation based on a method proposed by Black (17) was described by Bell (18) using a skewed quadrupole peak as a reference shape. The correlation function was extremely simple, using integer coefficients, and was given by $C = -Y_0 + Y_1 + 2Y_2 + Y_3 - Y_5$ where Y_0 to Y_5 represent six successive digital samples. The coefficients were derived not only to provide correlation with the reference mass spectral peak but also to suppress the effect of background not associated with structure. The correlation function was then thresholded in the normal manner. Although this correlation function has some similarity to digital smoothing techniques, it should be noted that such smoothing techniques are of general applicability, whereas the correlation function described is designed to enhance one particular reference peak shape with respect to noise. The method described by Bell incorporated further tests. For example, the peak area was required to exceed a minimum value to reject noise peaks. A further requirement was that the intensity at any peak maximum should be at least twice that at the valleys 0·5 a.m.u. ahead and behind. This last test meant that some multiply charged ions would not be detected.

Computer techniques for improving the effective resolution of a MS may be applied to digital data (19, 20). The observed spectrum ($F(y)$) is the result of the convolution of the "true" distribution ($b(y)$) with an instrumental smearing function ($a(x - y)$) representing the limited resolving power of the instrument (eqn. 2.2). Solving eqn 2.2 to obtain an estimate of the "true" distribution is termed deconvolution:

$$F(y) = \int_{-\infty}^{\infty} a(x - y) b(y) \, dy. \tag{2.2}$$

In a practical test, Zabielski and McHugh (19) applied both iterative and Fourier techniques to the deconvolution of spectra from a defocused time of flight MS. The iterative technique successfully resolved all the peaks in the mass range chosen with adequate quantitative accuracy. Fourier methods, however, would not yield quantitative results for unresolved peaks differing by more than a factor of two or three in intensity. In either case, smoothing of the data was a necessary prelude to deconvolution.

2.2.10 Peak Position and Intensity Calculation

The next stage in data processing is the conversion of digitized data into a time and intensity measurement for each peak recognized. The two most common methods for estimating the time of a peak are (a) the centroid time and (b) the peak maximum time, in practice the time of the highest digital sample.

The centroid method (Fig. 2.8) is generally used where the accurate mass of a peak and therefore the accurate time is to be calculated from a succession of digital samples across that peak (21). For a smooth, symmetrical peak, the peak maximum time will coincide with that of the centroid. The deficiencies of the peak maximum measurement are that (a) at low intensities, when ion statistics cause the peak to be far from regular in shape (see Fig. 2.9), the peak maximum is a poor measure of peak position compared with the centroid and (b) even in cases where ion statistics are favourable and the peak maximum is a good measure of position, the finite sampling rate means that the time of the highest digital sample can be up to half a sampling interval distant from the peak maximum time. The use of the centroid as a measure reduces this error due to the finite sampling rate (21). Sources of error in accurate mass measurement are discussed more fully in Chapter 3.

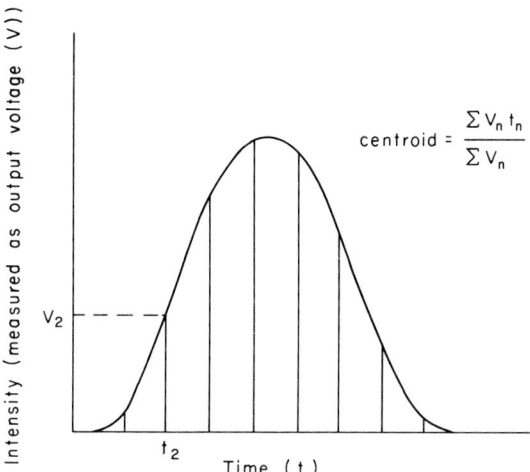

Fig. 2.8. Calculation of peak centroid from digital samples taken across the peak.

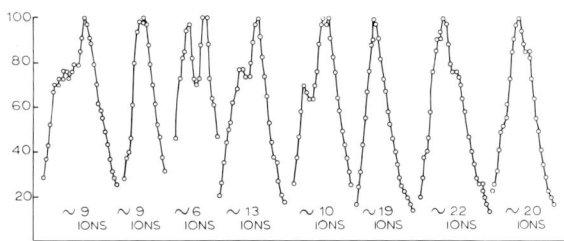

Fig. 2.9. Typical low intensity peaks from a digital print out. (From ref. 21.)

For nominal mass work, the accuracy of the method used to establish the peak position is less critical. Location of the peak maximum is computationally much simpler than centroid determination and therefore has been the method most commonly used (22), particularly when the computational load of centroiding would cause a reduction in the maximum scanning rate available for GCMS operation. However, as centroiding is a more accurate method, it can give more reliable nominal mass data under more demanding conditions such as the mass determination of weak peaks at high mass. Other methods for nominal mass peak location such as determination of the peak centre (10, 23) and a weighted average of the five most intense samples (24) have also been proposed.

The intensity of a peak is determined as either its maximum height or area. Peak height always gives a direct measure of intensity. Peak area is also a direct measure of intensity when the peaks have the same width in time at given percentage height, as in an exponential magnetic scan. McMurray *et al.* (3b) found that the standard deviation on area measurements was better than that on peak height and closely approached the theoretical deviation based on the number of ions in the peak. A disadvantage of area measurement is that it is, in general, a computationally more demanding method. However, in cases where the peak position is already being calculated by centroiding, area calculation becomes trivial as the necessary sum ΣV (see Fig. 2.8) has already been calculated. Other factors influence intensity measurements much more strongly than the choice of measurement method. These are factors such as the effect of changing source conditions particularly on the spectra of labile materials and the effect of changing sample concentration during a scan as a sample elutes from a gas chromatograph (GC). If the TIC is also digitized at suitably frequent intervals, this data can be used to correct mass spectral peak intensities for changes in concentration, although the accuracy of such corrections is probably not very high. A more detailed discussion of the use of data systems in quantitative mass spectrometry is given in Chapter 8.

2.3 ALTERNATIVE DATA ACQUISITION SYSTEMS USING AN ANALOGUE TO DIGITAL CONVERTER

Two important alternative systems for data acquisition from scanning MSs using an A/D converter have been described. These are the Carrick system (25) and the incremental recording system described by Jansson *et al.* (4b, 4c, 26). A major difference between these systems and the conventional system previously described is that the A/D converter is not operated in a continuously sampling mode so that the data flow to the computer is greatly reduced. An outline schematic of the Carrick system (25) is shown in Fig. 2.10. Essentially, the Carrick system differentiates the MS output to locate peak maxima (see Fig. 2.11). The output of the logarithmic amplifier is fed in parallel to a S/H amplifier and a differentiating amplifier. A peak maximum corresponds to a falling differentiated signal passing through zero. At the time it passes through zero, the differentiated signal triggers the hold function of the S/H amplifier, thus retaining the peak maximum intensity value. The A/D converter is then triggered to sample this intensity value once. The time value, taken from a 1 MHz clock, corresponding to the differentiated signal passing through zero is also sampled and retained as the peak time. Immediately following digitization, time and intensity data is written into a buffer memory. At the end of a scan the digital mass spectral record can be transferred to digital magnetic tape or output on paper tape for later computer

DATA ACQUISITION

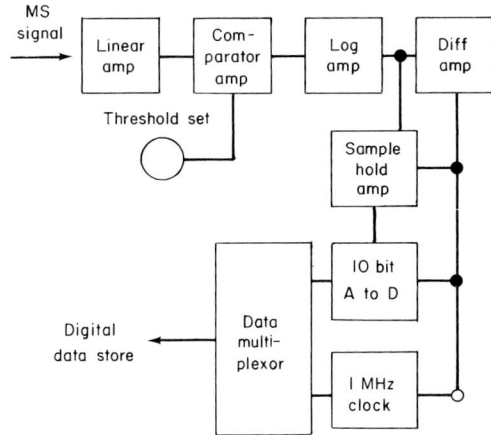

Fig. 2.10. Outline schematic of Carrick system. (Adapted from ref. 4a).

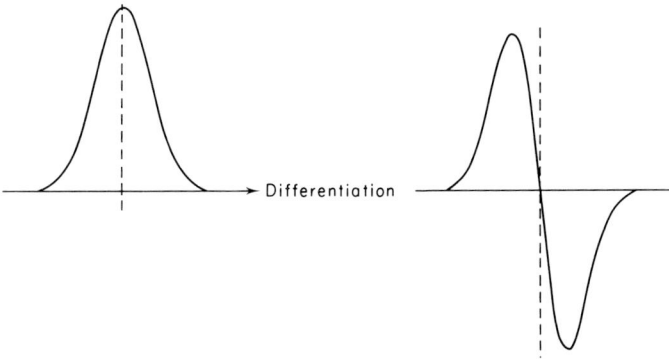

Fig. 2.11. Location of peak maximum by differentiation.

processing to mass and intensity data or it can be transferred directly on-line to a computer. The main virtue of the Carrick interface lies in its ability, in a relatively simple manner, to reduce the data flow for further processing. In on-line work, this means that the work load on the central processor is reduced, a particular advantage in fast scanning conditions. In off-line work, the storage capacity of a device such as digital magnetic tape is greatly increased. For example, Grayson et al. (4a) quote a $10 \times$ increase in the number of spectra that can be stored on a given length of tape compared with the conventional comparator and continuously sampling A/D converter system (cf. Table 2.1).

Weaker peaks, particularly when recorded at fast scanning speeds or at high resolution will present a ragged appearance which may result in the recording of more than one maximum with the Carrick system. This fact and the fact that, as previously mentioned, the peak maximum of weaker peaks is a poor measure of peak position compared with the centroid are disadvantages of this type of system, particularly when accurate mass measurements are required. In addition, no measurement of peak width is presented, thus, the presence of unresolved multiplets cannot be established, nor can noise spikes easily be rejected.

In the incremental recording system, a peak following amplifier holds the maximum peak intensity value which is subsequently read once by the A/D converter initiated by a pulse from a Hall probe or other form of mass marker (26). The system is set so that the A/D converter reading is taken about 0·3 a.m.u. after the calibrated integer mass (Fig. 2.12).

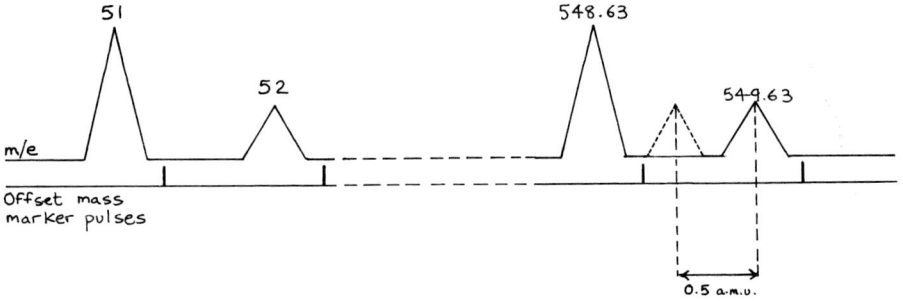

Fig. 2.12. Operation of incremental recording system with mass marker calibrated for saturated hydrocarbons. (From ref. 26.)

The incremental system is a nominal mass only system, but is simple to construct and produces only one intensity value at each mass, considerably reducing the amount of computer processing required. However, the system can give a somewhat less reliable mass scale than a system not based on a mass marker since no correction for changes in the fractional mass of peaks based on recorded peak data can be made (cf. Section 3.2.2). For example, if the mass marker is calibrated for standard hydrocarbons, peaks over a fractional mass range of about 0·5 a.m.u., depending on the resolving power, can be accepted, as shown in Figs 2.12 and 2.13.

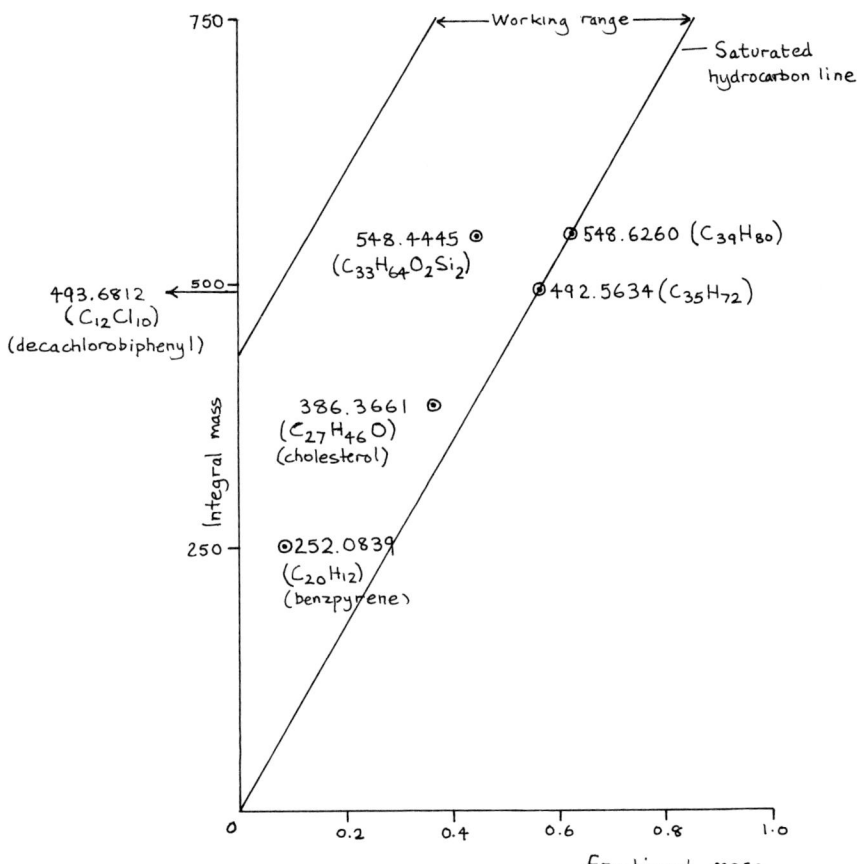

Fig. 2.13. Working range of incremental recording system. Mass marker calibrated for saturated hydrocarbons. (Adapted from ref. 26.)

Most organic compounds are found within this range. However, for the correct registration of very mass deficient compounds such as polyhalogenated compounds a resetting of the calibration is required. The problem becomes more acute at higher masses as the possible range of fractional masses increases.

2.4 DATA ACQUISITION FROM QUADRUPOLE MASS SPECTROMETERS

Most of the preceding discussion of data acquisition from scanning MSs is equally applicable to magnetic sector or quadrupole instruments. However, because of the low voltages involved in their operation, quadrupole mass filters are much more amenable to computer control. For example, the voltage applied to the quadrupole filter rods may be rapidly scanned with the smallest step of the D/A converter (see Fig. 1.7) that is available. With a 14 bit D/A converter, this would mean a division of the scan into 16 383 steps, rather less than 0·1 a.m.u/step for a mass range of 1000 a.m.u. If the MS amplifier output is sampled by an A/D converter at each step, the resulting data is analogous to the digitized data from the scan of a magnetic sector instrument. However, an alternative system is possible in which the mass filter is "stepped" under data system control directly from one integer mass to the next, the voltages corresponding to each mass position having previously been established by a calibration run (27). Like the Carrick system and the incremental recording system, this method produces only one intensity value for each mass position and is computationally undemanding. A further advantage of this method is that such a data collection system may be set to integrate the signal at each mass for a specified length of time before the output is sampled by the A/D converter. Thus, at higher mass, where weaker peaks may be expected, the output may be integrated for a longer period to improve the S/N ratio and therefore peak detectability (27). A version of this system has been developed that establishes integration times on-line as a function of signal strength so that an approach to optimum sampling conditions is obtained (28).

A drawback of this mass to mass stepping system is that a deviation between the fractional mass of the sample peaks and the fractional mass of the reference peaks used in calibration, can result in erroneous sample peak intensity measurements. In addition, particularly at high mass, assignment of an incorrect nominal mass becomes possible under these conditions. These problems are further compounded by any drift in the calibration function with time. Perhaps the most fruitful application of the stepping system is in the quantitative determination of specific compounds by stepping between a limited number of selected masses. This application is considered in more detail in Chapter 8.

2.5 DATA ACQUISITION FROM FOCAL PLANE MASS SPECTROMETERS

The Mattauch–Herzog geometry differs from others in that the whole mass range is brought into focus at the same time. The points of focus of all the masses lie on a plane, so that the device that has been used generally to record spectra in these instruments is the photographic plate (29). Ions incident on the emulsion cause the formation of latent images of silver specks which can be chemically developed to provide visible blackening of the plate as a measure of ion beam intensity. Distance and density information on the photoplate is measured by means of a microdensitometer and then transmitted to the data system.

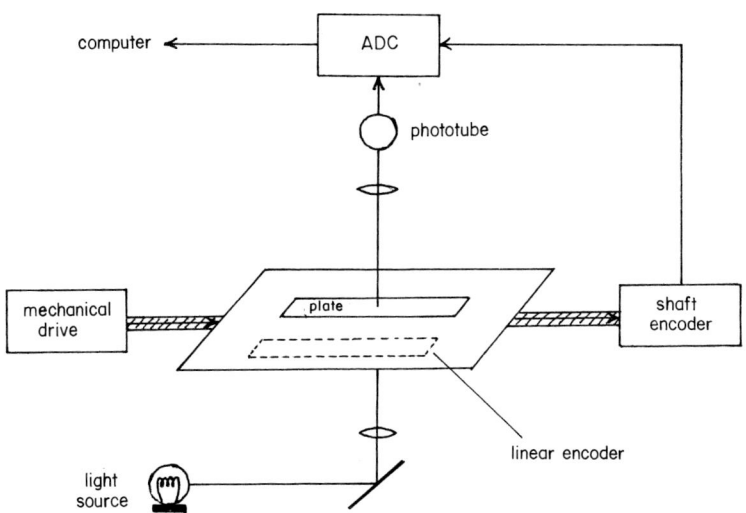

Fig. 2.14. Schematic of automated microdensitometer system.

An overall schematic of such a system is shown in Fig. 2.14 (29). The photoplate is mounted on the carriage of the microdensitometer and the carriage moved, along the length of the recorded spectrum by means of a motor whose motion is transmitted to the plate via a very finely threaded, hardened steel screw. The position of the plate at any instant can be derived from pulses originating from a shaft encoder mounted on the axis of the screw. The screw has been accurately machined, usually so that each rotation corresponds to precisely 1 mm. Thus with 1000 encoder pulses/rev, each pulse corresponds to a motion of the carriage by 1 µm. Each pulse also triggers the reading of a data point from the plate. In an alternative system, the position is determined from pulses derived from a linear encoder which is fixed on the carriage together with the plate and therefore travels with it to give a direct indication of position (30). The basic system used to determine the optical density of the lines on the plate can also be seen in Fig. 2.14. A light beam, obtained from a stable source and defined to be of dimensions slightly larger than those of a line, illuminates the plate, usually from underneath. The light beam passing through the plate is detected and amplified by a photomultiplier and the resultant signal digitized by an A/D converter.

In high resolution organic mass spectrometry, where the accurate measurement of line position is most important, data is usually read at $\frac{1}{2}$ μm intervals. In inorganic spark source mass spectrometry, when ion abundance is the most important measurement, data is read at approx. 5 μm intervals. In either case, the time needed to read a plate in most systems is 5–10 min, corresponding to a scanning speed of 1–0·5 mm s^{-1}. However, descriptions of two systems intended for inorganic work have appeared more recently in which a scanning speed of 1 cm s^{-1} has been used (31). Since treatment of the photoplate data is an off-line process and because the plate has stored all the original spectral information, the digitized data can be subjected to more complex processing, e.g. line recognition and deconvolution procedures, than would be appropriate or possible with electrically recorded data from a scanning MS.

A description of the procedure used in processing high resolution photoplate data to give line positions and intensities is given by Biemann (32). This procedure, applied to the digitized data, comprises the following basic steps:

(a) Calculate the average and standard deviation (σ) of points not representing a line (i.e. peak).
(b) Set the threshold 5σ above the average background.
(c) Calculate all peak centres above the threshold using the centroid as well as the three highest points.
(d) Deconvolute if the two values for the centre do not agree within specified limits.
(e) Store each peak (position of centre and integrated density). The optical density values are later converted to intensity values using xenon as an internal standard to eliminate the need for reproducible development of the plates.

Other systems described for use in high resolution photoplate work employ curve fitting to a Gaussian model (33).

A very similar system to that described by Biemann has been used by Frisch and Reuter (34) for identification of peaks and determination of their position on plates from inorganic spark source analysis. Using voltage readings V, which represent optical transmission and therefore have a minimum value at the peak top, a search is made for all sets of five or more consecutive points for which $V(2) < V(1)$ and where no points exist for which $V(N + 1) < V(N) > V(N - 1)$, in order to exclude closely spaced multiple noise peaks. A second order least squares fit is made to the remaining points and this then constitutes the background above which peaks are now reidentified by a method similar to that of Biemann (Steps (b) and (c), p. 42).

Most systems also employ a 5 point smoothing procedure on the raw data before any further processing is carried out. An elegant variation of this procedure has been used by Millett *et al.* (30) in their spark source system. By use of the convoluting integers determined by Savitzky and Golay (9) to fit a first derivative curve rather than a curve from the original data, the onset of a peak may be detected when the slope of the first derivative curve rises above a threshold value. In the same way, the turning point and end of a peak may also be determined using this function. For calculation of the peak position, some form of curve fitting to a limited number of points around the intensity maximum followed by location of the maximum of this fitted curve is often preferred. It has been suggested (34) that when, as in spark source operation, the peaks are more markedly asymmetric, this procedure gives a better estimate of peak position than do computations using the whole peak profile, e.g. centroiding or curve-fitting of the whole peak profile. Asymmetric peaks can, of course, also arise because of the presence of unresolved multiplets and the method of Biemann (Step (d), p. 42) uses the exaggerated difference between the centroid and the profile maximum to indicate the presence of such a multiplet. If the values are significantly different (e.g. $\gtrsim 0.75\,\mu m$), then the profile probably represents a multiplet and can be deconvoluted by one of a number of methods. The procedures used to convert digital photoplate data to ion intensity data in spark source analysis are described in Section 8.10.

Photoplates have a number of disadvantages as ion detection systems. The principal disadvantages are the lack of a simple linear relationship between blackening and ion intensity necessitating a prior calibration of the emulsion and the time-consuming nature of data reduction from a photoplate. However, a new focal plane detector, recently described, promises much greater convenience of operation and a one-to-one correspondence between incident ions and the output signal, as well as a greatly increased sensitivity for the recording of a complete mass spectrum (35–37). This device is based on the use of a microchannel electron multiplier array in which each component of the array can have a diameter as small as $10\,\mu m$. The array is placed in the focal plane of the MS to act as an ion-electron converter together with a phosphor screen which converts the amplified electron line image into a light image of the mass spectrum. This light image is then available for transmission to a suitable recording device, such as a vidicon camera (38), from which a data system may acquire the spectrum.

REFERENCES

1. Klopfenstein C. E. (1974). *J. agric. Fd Chem.* **22**, 736.
2. (a) Johnson M. W., Gordon B. W. and Self R. (1973). *Lab. Pract.* **22**, 267.
   (b) Black P. J. and Cohen A. I. (1972). 20th Annual Conference on Mass Spectrometry and Allied Topics, Dallas. Paper R12, p. 380.
3. (a) McMurray W. J., Green B. N. and Lipsky S. R. (1966). *Analyt. Chem.* **38**, 1194.
   (b) McMurray W. J., Lipsky S. R. and Green B. N. (1968). *In* "Advances in Mass Spectrometry" (Ed. Kendrick E.), Vol. 4, p. 77. Institute of Petroleum, London.
4. (a) Grayson M. A., Putnam J. M. and Yaeger F. J. (1976). *Int. J. Mass Spectrom. Ion Phys.* **22**, 365.
   (b) Hedfjäll B., Jansson P. A., Marde Y., Ryhage R. and Wikström S. (1969). *J. Phys.* (E), **2**, 1031.
   (c) Jansson P. A., Melkersson S., Ryhage R. and Wikström S. (1969). *Ark. Kemi* **31**, 565.
5. (a) Brand M. J. D. and Fleet B. (1969). *Chem. in Britain* **5**, 557.
   (b) Perone S. P. and Jones D. O. (1973). "Digital Computers in Scientific Instrumentation". McGraw-Hill, New York.
6. Banner A. E. (1966). *J. scient. Instrum.* **43**, 138.
7. Birkner D. A. and Fishel D. L. (1974). 22nd Annual Conference on Mass Spectrometry and Allied Topics, Philadelphia. Paper D7, p. 130.
8. (a) Yinon J., Ashkenazi Y., Gilath Ch., Meri M., Grinberg M. R. and Wolf D. (1972). *Chem. Instrum.* **4**, 255.
   (b) Kögler W. (1971). *Chem. Anlagen Verfahren.* **81**, 76.
9. Savitzky A. and Golay M. J. E. (1964). *Analyt. Chem.* **36**, 1627.
10. Hites R. A. and Biemann K. (1967). *Analyt. Chem.* **39**, 965.
11. Weichert D. H., Russell R. D. and Blenkinsop J. (1967). *Can. J. Phys.* **45**, 2609.
12. Porchet J. P. and Gunthard Hs. H. (1970). *J. Phys.* (E) **3**, 261.
13. Edwards T. H. and Willson P. D. (1974). *Appl. Spectrosc.* **28**, 541.
14. Holmes W. F., Holland W. H. and Parker J. A. (1971). *Analyt. Chem.* **43**, 1806.
15. Hedfjäll B. and Ryhage R (1975). *Analyt. Chem.* **47**, 666.
16. Schwarzenbach R. and Clerc J. T. (1973). *Org. Mass Spectrom.* **7**, 1215.
17. Black W. W. (1969). *Nucl. Inst. Meth.* **71**, 317.
18. Bell N. W. (1973). 21st Annual Conference on Mass Spectrometry and Allied Topics, San Francisco. Paper K2, p. 193.
19. Zabielski M. F. and McHugh T. M. 21st Annual Conference on Mass Spectrometry and Allied Topics, San Francisco. Paper K4, p. 198.
20. Dromey R. G. and Morrison J. D. (1971). *Int. J. Mass Spec. Ion Phys.* **6**, 253.
21. Halliday J. S. (1968). *In* "Advances in Mass Spectrometry" (Ed. Kendrick E.), Vol. 4, p. 239. Institute of Petroleum, London.
22. Binks R., Cleaver R. L., Littler J. S. and MacMillan J. (1971). *Chem. in Britain* **7**, 8.
23. Plattner J. R. and Markey S. P. (1971). *Org. Mass Spectrom.* **5**, 463.
24. Habfast K., Carter Cook J. and Rinehart K. L. (1972). *In* "Biochemical Applications of Mass Spectrometry". (Ed. Waller G. R.), p. 121. John Wiley and Sons, Chichester.
25. Carrick A. (1969). *Int. J. Mass Spectrom. Ion Phys.* **2**, 333.

26. Hedfjäll B., Ryhage R. and Åkerlind Å. (1974). *In* "Advances in Mass Spectrometry" (Ed. West A. R.). Vol. 6, p. 1043. Applied Science Publishers, Barking.
27. Reynolds W. E., Bacon V. A., Bridges J. C., Coburn T. C., Halpern B., Lederberg J., Levinthal E. C., Steed E. and Tucker R. B. (1970). *Analyt. Chem.* **42**, 1122.
28. Eichelberger J. E. and Budde W. L. (1974). 22nd Annual Conference on Mass Spectrometry and Allied Topics, Philadelphia. Paper D6, p. 129.
29. Desiderio D. M. (1971). *In* "Mass Spectrometry, Techniques and Applications" (Ed. Milne G. W. A.), p. 11. John Wiley and Sons, Chichester.
30. Millet E. J., Morice J. A. and Clegg J. B. (1974). *Int. J. Mass Spectrom. Ion Phys.* **13**, 1.
31. (a) Witmer A. W. and van Gool G. H. (1971). Colloquium Spectroscopicum Internationale XVI, Heidelberg. Reprints V2, Paper 56, p. 254. Adam Hilger, London.
    (b) Franzen J., Schönfeld W. and Stüwer D. (1971). *In* "Advances in Mass Spectrometry" (Ed. Quayle A.), Vol. 5, p. 322. Institute of Petroleum, London.
32. Biemann K. (1972). *In* "Biochemical Applications of Mass Spectrometry" (Ed. Waller G. R.), p. 96. John Wiley and Sons, Chichester.
33. (a) Venkataraghavan R., McLafferty F. W. and Amy J. W. (1967). *Analyt. Chem.* **39**, 178.
    (b) Tunnicliff D. D. and Wadsworth P. A. (1968). *Analyt. Chem.* **40**, 1826.
34. Frisch M. A. and Reuter W. (1973). *Analyt. Chem.* **45**, 1889.
35. Giffin C. E., Boettger H. G. and Norris D. D. (1974). *Int. J. Mass Spectrom. Ion Phys.* **15**, 437.
36. Tuithof H. H. and Boerboom A. J. H. (1976). *Int. J. Mass Spectrom. Ion Phys.* **20**, 107.
37. Tuithof H. H., Boerboom A. J. H. and Meuzelaar H. L. C. (1975). *Int. J. Mass Spectrom. Ion Phys.* **17**, 299.
38. Talmi Y. (1975). *Analyt. Chem.* **47**, 699A.

3

Data Conversion

3.1 INTRODUCTION

The first requirement for the conversion of peak position data into mass data is to establish a calibration. In accurate mass determination, this calibration is a correlation between the time centroid and mass of peaks of known composition from a standard compound such as perfluorokerosene (PFK). To provide sufficiently accurate mass measurement, the standard is run concurrently with the sample, usually admixed with it. In nominal mass determination (1, 2, 3, 7, 10b, 11), the variable correlated with mass may be something other than time. For example, quadrupole mass spectrometers are calibrated through a correlation between quadrupole rod voltage and mass whilst magnetic sector instruments using a Hall probe mass marker use a correlation between Hall voltage and mass. As in accurate mass determination, the peaks of known mass originate from a reference compound such as PFK or heptacosafluorotributylamine. However, because calibrant peaks are usually not resolved from sample peaks at the same nominal mass under low resolution conditions, nominal mass calibration uses an external standard, run separately from the sample.

3.2 NOMINAL MASS DETERMINATION

3.2.1 Calibration

Having acquired data from a scan of the calibration compound, it is initially necessary to recognize one or two major peaks as starting points for the calibration. Automatic recognition of one or two major peaks of the calibrant (e.g. m/e 69 and 119 in PFK) or of major background peaks (m/e 28) can be made relatively easy by ensuring that sufficient calibrant or air (introduced inadvertently or otherwise) enters the source to make these peaks the biggest in the spectrum. The most likely candidates for these peaks are chosen from the calibration scan on the basis of their intensities and approximate position in the scan. Confirmation that these peaks have been chosen correctly may be made by relative intensity checks or preferably by using these peaks and a knowledge of the scan law to predict, by extrapolation, the position of further major calibrant peaks. A search within a small tolerance around the predicted position should then locate the expected calibrant peaks. Prediction and location of all major calibrant peaks, whose masses are stored in the computer, is made in the same way. Should this prediction and location procedure fail, the next most likely candidate peak for the starting mass would be chosen and the procedure repeated.

A successful calibration can then be summarized in one of two ways:

(a) For scans that are a reasonable fit to a simple scan law—by constants in a scan law equation. For example, the constant k in eqn 3.1 for a time of flight scan that always starts at the same driving voltage. Also, the constants a and b in the equation $m = aV + b$ for a quadrupole mass filter.

(b) For less regular or less simply described scans—as a lookup table. A lookup table might consist of times of all major calibrant masses or of all integer masses or of time intervals between adjacent integer masses. An important consideration in construction of these tables is to make subsequent interpolation using the tables to give sample masses, as simple as possible. Thus, linear interpolation is to be preferred wherever it yields sufficiently accurate results for nominal mass operation, e.g. between adjacent integer masses or between values of some function of mass that varies more linearly with time (1):

$$m_2 = \left[m_1^{1/2} + \frac{t_2 - t_1}{k} \right]^2. \tag{3.1}$$

3.2.2 Determination of Mass from Peak Times—Rounding off to Nominal Mass

Sample masses are determined by insertion of time or other data in the appropriate scan equation or by interpolation between calibrant data in the lookup table. Although the data finally output is usually only the nominal mass, it is, however, necessary initially to determine the mass more accurately than this. This is partly because an approximate knowledge of the fractional mass can be useful in determining compound type, but also because a reasonable knowledge of the fractional mass is necessary to be able to achieve a rounding off to the correct nominal mass.

The task of rounding off a fractional mass to the corresponding nominal mass is more complex than is at first apparent. For example, the saturated hydrocarbon $C_{20}H_{42}$ with a fractional mass of 282·33 should be rounded off to 282 and the hydrocarbon $C_{40}H_{82}$ with a fractional mass of 562.64 should be rounded off to 562. However, the chlorinated biphenyl, $C_{12}Cl_{10}$, with a fractional mass of 493·69 (taking chlorine as ^{35}Cl) should be rounded off to 494. Rounding off is usually accomplished using a correction procedure due to Hites and Biemann (2). This procedure relies on the fact that although the fractional mass varies with total mass, its rate of change throughout the spectrum of a single compound is relatively constant, although, of course, admixtures of very different fractional mass can cause problems.

A version of this correction procedure is summarized in Table 3.1 and in the following discussion.

TABLE 3.1
Hites–Biemann correction procedure

Fractional mass (calculated from corrected time T)	Action 1	Action 2
0·9—0·1	Update ratio	Print as nominal mass
0·1—0·4	Update ratio	Correct this and all subsequent times
0·4—0·6	No action	Print as half mass
0·6—0·9	Update ratio	Correct this and all subsequent times

The correction routine maintains an average ratio Δ (the average diminishes the effect of an occasional peak of widely different fractional mass) given by eqn 3.2 for the last five peaks that were above 0·1 % of the base peak and which were not half masses. In this equation, T is the measured peak time, T_{meas}, corrected according to eqn 3.5 and T_{nom} is the time of the nearest nominal mass. The ratio, Δ, is used, as shown in Table 3.1, to correct the present peak time according to eqn 3.3 and all subsequent peak times according to eqns 3.4 and 3.5:

$$\Delta = \sum_{n=1}^{5} \left(\frac{T_{nom}}{T}\right)_n \bigg/ 5 \qquad (3.2)$$

$$T = T \times \Delta \qquad (3.3)$$

$$(\text{correction factor}) = (\text{correction factor}) \times \Delta \qquad (3.4)$$

$$T = T_{meas} \times (\text{correction factor}). \qquad (3.5)$$

This correction procedure is re-established and the correction factor re-set at 1·0 for each scan. Use of the Hites–Biemann correction overcomes two problems; (a) the gradual divergence of true masses from a nominal value through the mass scale and (b) any change with time in the mass-time or mass-voltage relationship from the calibration scan to the sample scans.

In effect, the Hites–Biemann correction uses the sample itself as a calibrant to provide corrections. A number of other correction procedures using sample peaks have been proposed. One of these is the method due to Binks et al. (3). In this method, the lookup table consists of the time intervals between peaks one mass unit apart at all m/e values (see Fig. 3.1). These time intervals are appreciably different for each mass interval at low mass because the scan used is an exponential magnet scan from high to low mass. Using the lookup table, each sample peak identified in turn serves as a reference point to find the next higher mass peak. Thus, if sample peak #10 (Table 3.2) has been identified as m/e 77, the time of sample peak #11 is subtracted to give 4304 time units and then compared with summed time intervals from 77 taken from the lookup table (viz. 1104, 2178, 3243, 4299), until the two figures are almost equal. Having identified peak #11 as m/e 81, the time

LOOKUP TABLE
10/8 /70
R6 D9

	0	1	2	3	4	5	6	7	8	9
0	0	0	0	0	0	0	0	0	0	0
10	0	0	0	·0	0	5557	5275	5045	4631	4433
20	4244	4075	3737	3612	3473	3362	3257	3161	3057	2771
30	2707	2632	2556	2505	2435	2401	2336	2275	2245	2207
40	2157	2124	2052	2022	1774	1746	1720	1674	1650	1625
50	1603	1562	1541	1521	1501	1462	1444	1426	1410	1373
60	1356	1342	1326	1313	1300	1265	1253	1241	1227	1215
70	1206	1174	1164	1153	1143	1133	1123	1113	1104	1074
80	1065	1056	1050	1041	1033	1024	1016	1010	1002	774
90	767	761	754	747	742	735	730	723	716	711
100	705	700	674	670	663	657	653	647	643	640
⋮										
570	125	125	125	124	124	124	124	124	124	124
580	123	123	123	123	123	123	123	122	122	122
590	122	122	122	122	121	121	121	121	121	121
600	121	121	120	120	120	120	120	120	120	117
610	117	117	117	117						

Fig. 3.1. Lookup table. (From ref. 3.)

interval sum is re-set to zero and the process repeated for the next peak. Unlike a method using times from the beginning of the scan, the accumulated mass defect error at each stage is not large so that a satisfactory match with the nominal mass time intervals can usually be obtained. Peaks within a set tolerance of the summed time interval are assigned a nominal mass; those which cannot be so assigned are checked within a similar tolerance to see if they could be half masses, if not they are left unassigned. The system is computationally simple but is only applicable where the scan law results in an interval between adjacent masses that is not constant throughout the scan. The system described by Binks et al. is also advantageous in that use of

TABLE 3.2
Correlation of peak time and mass using lookup table in Fig. 3.1

Peak number	Peak time	m/e value assigned
10	117324	77
11	113020	81
12	110935	83

time intervals means that its operation is not so critically dependent on reproduction of start position as is the peak time lookup table method. It is also reported that time intervals are very stable and such a calibration can hold for several months.

Another method that uses sample peaks for calibration has been described by Grayson and Conrads (4). This method has been applied to a time of flight MS, in which the mass of an ion may be calculated from its flight time using eqn 3.1. The time base used is the gate driving voltage from the analogue scanner used with the time of flight MS (see Section 1.2.5). Since there is a fixed correspondence between each value of this voltage and each point on the mass scale, use of this voltage as the time base makes the procedure independent of scan speed and also less susceptible to small irregularities in the scan. The system is calibrated with benzene, although any compound whose most intense peak is unequivocally known can be used. From the value of the driving voltage at the scan start (m/e 14) and at the base peak (m/e 78 in benzene) a value for k is calculated. This is the only data stored from the calibration. This value of k is used in eqn 3.1 to calculate the mass of the first peak in the sample spectrum by reference to the scan start and then that of the second peak in the same way. The third peak is calculated by reference to the first peak, the fourth from the second and so on, but always maintaining the same value of k. The continued shifting of the reference point (m_1, t_1) is a simple expedient to avoid calculation over large mass gaps for which eqn 3.1 is not strictly accurate, although the authors have achieved success with gaps of nearly 150 a.m.u. All calculated masses which fall within the round off limits of $M \pm 0.3$ a.m.u. are rounded off to the nearest integer mass, M. Any values falling outside these limits are disregarded. A similar system in which the computer calculates a mass scale from a given scan function of a magnetic sector MS, has been briefly mentioned (5).

The method of Grayson and Conrads does also constitute a method by which sample masses higher than those of the calibrant can be determined. As operation above m/e 1000 is not well catered for by reference compounds*, but is well within the capabilities of many magnetic sector instruments as well as time of flight instruments, a system of extrapolation to and identification of sample masses above m/e 1000 using existing reference compounds would be very valuable. An alternative suggestion for the determination of nominal mass values above m/e 1000 has been made by Plattner and Markey (1). After a normal calibration run with PFK (upper mass limit m/e 700–900), the high mol. wt sample is analysed, at a sufficiently low accelerating voltage to cover the mass range needed, together with the PFK in the source. This mixed sample scan is then interpolated in the normal manner against the calibration scan. If, for example, the accelerating voltage has been lowered by a factor 2, then all the interpolated masses will be approximately a factor 2 too low. An accurate multiplication factor is then derived by identifying a PFK peak in the low accelerating voltage spectrum and dividing its assigned mass into its correct mass. This factor is then used to multiply each assigned mass value to obtain correct values. The factor may be recalculated for higher PFK peaks to give increased accuracy as the mass scale is ascended. Using this technique, an accuracy of 500 p.p.m. has been obtained as high as m/e 3600 (e.g. $\pm 1\cdot 5$ a.m.u. at 3628) although Plattner and Markey suggest m/e 2000 as a practical limit with their instrumentation because of sensitivity and resolution losses as the accelerating voltage is lowered.

3.2.3 Determination of Mass using Variables Other than Time

The system of Grayson and Conrads (p. 50) is one example of a system where a variable other than time is correlated with mass in the calibration. Other examples are the use of the voltage output of a Hall probe in magnetic sector instruments or the rod voltage in quadrupole mass filters. When a Hall probe, or other suitable mass marker, that gives a four digit display of the mass is available, the output of this display may be sampled at the same time as the beam data. Thus, a mass scale is imposed on the beam data and it is then a relatively simple matter for the software to assign nominal masses to peaks using this scale (6). The system described by Sweeley *et al.* (7) also uses a Hall probe, but without the mass display facility. In this case, calibration is used to provide a lookup table of reference masses and the corresponding Hall effect voltages. A set of coefficients to allow assignation of masses to

---

* Fomblin® oil (Montedison, Milan) has recently been found to provide an excellent reference spectrum covering a mass range up to at least m/e 3000.

sample peaks from the Hall effect voltage by interpolation between the values in the lookup table is also provided. These calculated exact sample masses are then converted to nominal masses with an updating procedure that corrects for mass defects as well as machine drift.

The calibration of a quadrupole MS is based on a scan of a reference compound such as heptacosafluorotributylamine. This scan is taken using small steps of the D/A converter controlling the rod voltages so that the position of each reference peak, usually measured as the peak top position, can be accurately determined. Thus, a calibration of the known reference masses against rod voltage is established. With this calibration, sample masses may generally be determined from the observed voltages by linear interpolation, although a slightly more complex procedure was adopted by Houseman and Hafner (8) for operation below m/e 14. The calibration may also be used to determine the appropriate rod voltages when the quadrupole is directed to step and sample (or step and sweep when drift is a problem (8)) rather than effecting a complete scan.

3.2.4 Instrumental Stability

Since an internal standard is not employed in low resolution operation, any lack of reproducibility in the scan can obviously lead to errors in mass assignment. A procedure that can be used to overcome the effect of drift in the mass-time or mass-voltage relationship is the use of a "lock-mass". For example, if an air peak (m/e 28 or 32) of reasonable intensity is present and can be unambiguously identified in the sample scans, the correction needed to give this peak its true mass can then be applied throughout the mass scale. Again, the use of the sample itself as a calibrant, as in the Hites–Biemann correction, provides a means of overcoming some variability in the scan law.

In a double beam MS (9), two ion sources are provided so that sample and reference compound may be ionized simultaneously but entirely independently. The second, or reference beam, then passes through the same analyser system as the first or sample beam, to provide a reference which is almost as effective as the use of an internal standard at high resolution. For this reason, the use of a double beam MS can provide accurate mass measurements at low resolving power as will be seen in the following section (p. 63). However, in addition, the double beam MS provides a highly reliable system for nominal mass determination.

3.3 ACCURATE MASS MEASUREMENT

The great majority of MS computer systems for accurate mass measurement (10, 12–15) have been based on the use of double-focusing instruments of

Nier–Johnson geometry, operating at high resolving power, usually 10 000. The principal steps in processing data acquired from such a system are (a) identification of all major reference peaks in a calibration scan, usually of PFK, run at high resolving power, (b) location of the same major peaks in a scan of a mixture of PFK and sample also run at high resolving power, and (c) interpolation between these reference peaks to determine the masses of the sample peaks.

3.3.1 Calibration

Identification of the peaks in the calibration scan begins with recognition of the base peak of the spectrum (m/e 69), usually the most intense peak in the scan. From the position of this and another intense reference peak, it is then possible to predict, by extrapolation, the position of all the other major reference peaks based on a knowledge of the scan law (cf. nominal mass calibration, p. 47). A search within a small tolerance around the predicted position is then used to locate each successive reference peak.

Sample scans usually start from the same magnet current and use the same scan speed as the calibration scan. The starting current is chosen so that the starting mass is well above the highest sample mass. Thus, the first part of the scan is purely PFK and the highest mass calibration peak can be sought as the most intense peak within a tolerance of the expected time for this peak. Having provisionally located this peak, the difference between its time in the sample run and the time of the corresponding peak in the calibration run is computed and stored as the "drift". A search is then made for the next two reference peaks at lower mass. These peaks are searched for at their calibration times, corrected by the drift and using a much smaller search tolerance, typically one peak width. Again the largest peak in each window is chosen as the candidate. These three peaks are then used to begin an extrapolation procedure to locate all the other major reference peaks. Another procedure relies on the location of m/e 69 (the base peak of PFK) as the largest peak within a certain time region of the sample scan, previously determined from the calibration scan. Extrapolation is then made from this and adjacent reference peaks through the whole mass range.

In some procedures, the most mass deficient peak within the search tolerance is chosen as the reference peak. Sample peaks are, in fact, usually less mass deficient, i.e. at a higher fractional mass, than those due to PFK, and sample levels are usually such that sample peaks are less intense than reference peaks at the same nominal mass. However, examples can arise where neither is the case. Thus, the spectrum of dieldrin shows an intense peak at m/e 343 with the composition $C_{12}H_8^{35}Cl_5O$ (m/e 342·9018). The major PFK peak at this same nominal mass has the composition C_8F_{13} (m/e 342·9796). Therefore, the more reproducible the scan and the more accurately the extrapolation procedure can predict the position of the next reference peak, the more likely it is to succeed (cf. double beam mass measurement, p. 63).

The most commonly used formulae for extrapolation in exponential scans are those due to McMurray et al. (eqn 3.6) (12). With this equation, given the times and masses of three consecutive major reference peaks at high mass (T_1, M_1), (T_2, M_2) and (T_3, M_3), a prediction is made of the time of the next reference mass M_4 as:

$$T_4 = T_3 + \tau_{34}(\ln M_3 - \ln M_4) \tag{3.6}$$

where

$$\tau_{34} = \tau_{23} + (\tau_{23} - \tau_{12})(M_2 - M_4)/(M_1 - M_3)$$

and

$$\tau_{12} = (T_1 - T_2)/(\ln M_2 - \ln M_1)$$

and

$$\tau_{23} = (T_2 - T_3)/(\ln M_3 - \ln M_2)$$

τ is the exponent in the exponential decay equation, $M = M_0 e^{-\tau t}$.

The exponent τ_{12} applies to the mass range between M_1 and M_2 and the value of τ is assumed to vary linearly with mass (M). This is a reasonable assumption except in the first 100–150 mass units covered by a scan where the change of exponent is more complex. Another equation used by Bowen et al. (13), also allows for a change of exponent, although, in this case, the change of exponent need not necessarily be linear with mass (eqn 3.7):

$$M = M_0 \exp(aT + bT^2 + cT^3). \tag{3.7}$$

Both equations are used in the same way, i.e. prediction of peak time, location of the peak and then use of the observed time in further prediction. A few authors have used simpler extrapolation formulae for exponential scans (14) and a report of a least squares fit used in extrapolation of linear scan data has appeared (15).

It is possible that in some cases a better method than extrapolation on the sample scan might be to prepare a lookup table of reference peak times from the calibration run. Preliminary results indicate that the method, after correction for drift, can function at least as well in locating reference peaks in the sample run as the complex extrapolation formulae and would be much simpler computationally. Once all the major reference peaks have been located, these are then used for calculation of the masses of all the remaining peaks by interpolation. Although these peaks are mainly sample peaks, smaller reference peaks are also present. However, it is possible to eliminate these peaks by having a table of their accurate masses stored in memory for comparison. The choice of a method of interpolation is important since this influences the accuracy of mass measurement and therefore the confidence with which an elemental composition can be ascribed to sample peaks.

3.3.2 Determination of Accurate Masses by Interpolation

A number of interpolation formulae have been described for exponential scanning instruments. One of the earliest was that due to McMurray et al. (12) (eqn 3.8):

$$M_u = M_2 \exp\frac{(T_2 - T_u)}{\tau_u} \qquad (3.8)$$

where

$$\tau_u = \tau_{23} + (\tau_{12} - \tau_{23})\frac{(T_3 - T_u)}{(T_3 - T_1)}.$$

The variables are those defined in eqn 3.6 and the u refers to the unknown sample peak. It is further assumed that M_u lies between reference masses M_2 and M_3. As can be seen, these equations are based on the same assumption of a linear change of scan time constant with mass as eqn 3.6. Each interpolation by this procedure uses three reference points.

More general forms of interpolation, again using precise fitting of reference peaks were investigated by Bowen et al. (13). The method found to give the best interpolation accuracy used four reference peaks, M_0, M_1, M_2, M_3 and three equations of the form of eqn 3.9:

$$M_n = M_0 \exp\left(a(t_n - t_0) + b(t_n - t_0)^2 + c(t_n - t_0)^3\right) \qquad (3.9)$$

where M_n takes the values M_1, M_2 and M_3 to calculate the constants a, b and c. These constants were then used to interpolate for the unknown peak (see Fig. 3.2). A similar system, also using four reference peaks, was described by McLafferty et al. (14b).

In the same paper Bowen et al. (13), also investigated interpolation methods where equations were generated by a linear regression method. Thus, the fitting equation was not constrained to pass through any of the reference points but was the best least squares fit to them. In these investigations, the fitting of an exponential with cubic exponents (eqn 3.10) to six reference points gave the best interpolation accuracy and gave results which were

	No. of reference peaks used for interpolation	Standard deviation (p.p.m.)	
	3	6·2	
	4	4·9	
	6	5·5	(a)
	8	6·0	
	10	6·7	

Form of equation used for regression	No. of reference peaks used for interpolation	Standard deviation (p.p.m.)	
$M = A \exp(at + bt^2)$	4	5·1	
"	6	5·1	
"	8	7·5	
"	10	12·2	(b)
$M = A \exp(at + bt^2 + ct^3)$	6	4·3	
"	8	4·6	
"	10	5·4	

Fig. 3.2. (a) On-line interpolation accuracies based on precise fitting of reference peaks. (b) On-line interpolation accuracies based on linear regression methods. These results are the standard deviations of calculated masses from true masses for 300 interpolations derived from 12 spectra. They are quoted in p.p.m. as this gives a truer representation of the error at all masses. (From ref. 13.)

TABLE 3.3

Mass measurement errors (in p.p.m.) using least squares fit

Perchlorobutadiene (nominal mass)	Normal[a]	No. of cal.pts/No. of coefficients			
		7/3	7/5	5/2	5/4
223	−0·1	+0·1	−0·3	−1·3	+0·1
224	+0·6	+0·9	+0·3	−1·0	+0·8
225	−0·5	−0·1	−1·0	−2·6	−0·2
226	−0·3	+0·1	−0·8	−3·0	+0·1
227	−0·5	0·0	−0·9	−3·7	+0·1
228	−0·4	+0·1	−0·9	−4·2	+0·4
229	−0·7	−0·2	−1·3	−5·2	+0·1
230	−0·5	0·0	−1·1	−5·6	+0·6
231	−0·6	−0·2	−1·3	−6·4	+0·6

[a] "Normal" colum refers to an exact fit to the calibration points. (From ref. 17.)

somewhat better than those obtained by precise fitting to eqn 3.9 (see Fig. 3.2):

$$M = A \exp(at + bt^2 + ct^3). \tag{3.10}$$

An investigation into least squares fitting procedures by Burlingame *et al.* (17) also resulted in improved interpolation data using a very similar set of conditions, viz. cubic coefficients and seven reference points (see Table 3.3). These improved results using least squares fitting are not unexpected since the line generated is not constrained to pass through any of the reference points, each of which will be subject to a basic statistical inaccuracy.

There are a number of other references in the literature describing procedures for the determination of accurate masses from scan data, in some cases from scans other than exponential. However, all of these systems conform to the outline procedure described above, viz. extrapolation to predict and locate reference peaks and interpolation to calculate sample peak masses.

3.4 ERRORS OF MASS MEASUREMENT

Although inaccuracies in the interpolation method used can contribute to mass measurement errors, a number of other sources of error exist. It is therefore appropriate to list and discuss these. They fall into three categories (a) ion statistical errors; (b) instrumental errors arising in the MS; and (c) errors of measurement arising in the data system.

3.4.1 Ion Statistical Errors

For many purposes, the peaks in the output of a scanning MS can be closely approximated in profile by an isosceles triangle. This profile is the result of the convolution of a rectangular cross section ion beam (generated by the source slit) with a rectangular collector slit of width equal to or less than that of the ion beam. The height of this peak at any point is proportional to the number of ions falling on the detector in unit time at that point. If the number of ions being generated in unit time in the source is large, then the triangular shape will be quite faithfully reproduced. However, if it is small this number will be subject to statistical fluctuations which, as a fraction, are large and will result in the peak assuming a ragged appearance (Fig. 2.9). This distortion of the profile results in a statistical distribution of the centroid from which the mass is usually calculated about the true position, i.e. on repeat determinations, the centroid values will show a scatter about the true value. As the number of ions decreases so this scatter increases.

The expected standard deviation (σ) of centroid determinations on a peak is given by eqn 3.11:

$$\sigma = \frac{10^6}{R\sqrt{24N}} \tag{3.11}$$

where N is the number of ions in the peak and R is the resolving power used. For example, at 10 000 resolving power, with a peak containing only 10 ions and therefore near the limits of detection, $\sigma = 6\cdot 4$ p.p.m. Since the sample peak mass is calculated by interpolation, account must also be taken of the effect of statistical variation in the position of the reference peaks used for interpolation. The effect of reference peak variation may be illustrated by considering an approximation in which the unknown peak position is determined by linear interpolation between the two adjacent reference peaks (Fig. 3.3).

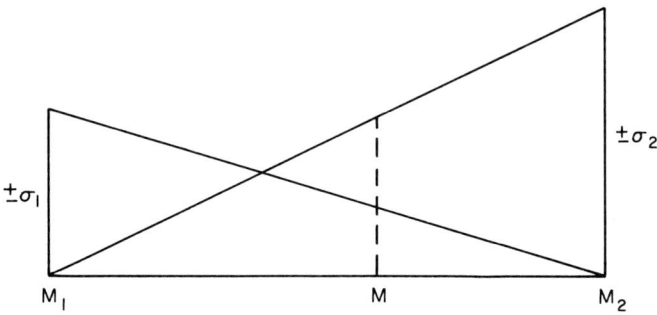

Fig. 3.3. Effect of reference peak ion statistics on interpolated data. Reference masses M_1 and M_2. Sample mass M. Respective standard deviations of reference peaks σ_1 and σ_2 and of sample mass σ.

The effect of σ_1 at M will be: $\sigma_1[(M - M_2)/(M_1 - M_2)]$, and the effect of σ_2 at M will be: $\sigma_2[1 - (M - M_2)/(M_1 - M_2)]$. Thus if $x = [(M - M_2)/(M_1 - M_2)]$, then:

$$\sigma_M^2 = x^2\sigma_1^2 + (1 - x)^2\sigma_2^2 \qquad (3.12)$$

and if we now include the statistics of the sample peak itself, then:

$$\sigma_M^2 = \sigma^2 + x^2\sigma_1^2 + (1 - x)^2\sigma_2^2. \qquad (3.13)$$

For example, consider a sample peak whose standard deviation is 5 p.p.m. due to ion statistics and which is measured by interpolation between two reference peaks whose standard deviations are each 5 p.p.m. If the sample peak is adjacent to one of the reference peaks, the resultant standard deviation is approx. 7·1 p.p.m. If the sample peak lies halfway between the two reference peaks, the standard deviation is approx. 6·1 p.p.m. Thus, the effect of error in the reference peaks used for interpolation is greatest when the sample peak is adjacent to one or other of these peaks and this effect will be especially evident if it is adjacent to one of the less intense PFK peaks, e.g. at high mass. However, if a least squares fit rather than exact curve fitting is used in calibration, the errors are no longer so localized and eqns 3.12 and 3.13 do not now apply.

3.4.2 Instrumental Errors

The voltage supplies that determine the mass focused at the collector, particularly the magnet power supply, can be affected by the pickup of small alternating currents (ripple). The result of these small periodic variations is that a focused peak will move from side to side across the collector slit and, in effect, vary in mass. Careful siting and shielding of components can reduce this type of pickup to acceptable levels. Movement of the peak may also result from mechanical vibration of the collector assembly, for example, due to vibration from rotary pumps or fan motors. Careful design of rotary pump connecting lines and gas chromatograph (GC) fan motor mountings to minimize transmission of vibration and perhaps a special vibration free-mounting for the whole instrument, are the usual steps taken to reduce this effect to acceptable levels.

Irregularities in the scan law are also a cause of mass measurement errors, although these are minimized as far as possible by the use of an internal standard and by the fitting of the most suitable equations relating mass and time. Obviously, these irregularities will vary from instrument to instrument and equations suited to one instrument may not suit another. However, two generally applicable points may be made. First, full cycling of the magnet one or two times before any data is recorded, reduces magnet hysteresis effects and ensures maximum reproducibility of subsequent scans. Secondly, scans tend to follow a more complex law near their commencement. Thus, for accurate mass measurement, a scan should be started at least 100–150 a.m.u. above the peaks to be measured.

3.4.3 Errors of Measurement and Calculation

An analysis may be made of the accuracy of digitization of output voltages. With 20 samples taken across a peak, the maximum voltage change between samples may be considered to come from a full scale peak and will therefore be approx. 10% full scale deflection (f.s.d.). During a 1 µs settling time, a sample and hold (S/H) amplifier is able to acquire this signal change to within 0·1%, i.e. 0·01% f.s.d. Other sources of error include the accuracy of the S/H amplifier ($\pm 0.025\%$ f.s.d.), the accuracy of the A/D converter ($\pm 0.01\%$ f.s.d.) and the resolution of the A/D converter of $\pm \frac{1}{2}$ least significant bit, i.e. $\pm 0.012\%$ f.s.d. for a 12 bit A/D converter. These errors combine to give an overall digitizing accuracy of $\pm 0.031\%$ f.s.d. i.e. ± 3.1 mV root mean square (RMS) for a full scale signal of 10 V.

The long term frequency stability of the timing source used is another possible source of error, but, in any case, errors introduced into the sampling periods which are the basis of mass measurement, are compensated for by means of the calibration with reference compound. In some experiments using an artificial signal consisting of a truncated 1 kHz sine wave to simulate peaks recorded under fast scanning, high resolution conditions, and sampling every 12 μs with a 13 bit A/D converter, Bowen et al. (13) demonstrated that centroids could be determined to better than 0.1 p.p.m. Such an error is less than any that might be expected on an ion statistical basis and therefore such a digitization system does not contribute materially to the overall mass measurement error.

Accuracy of mass measurement is also dependent on the accuracy with which centroid times, once calculated are kept, and on errors that are introduced during time to mass computations. Any variable expressed in binary form will have an absolute error of $\pm \frac{1}{2}$ least significant bit. Thus, with 23 bits, the relative error is 1 in 24 before any calculations are done. Further errors may be introduced during calculations, most particularly from the use of approximations, e.g. for exponentiation.

3.4.4 Effect of Digitization Frequency on Mass Measurement Accuracy

The choice of the correct digitization rate for mass measurement data has been treated in some detail by Halliday (18). The basic problem is that if the digitization rate is not fast enough, the small number of digital samples taken will be insufficient to describe the peak profile and consequently any subsequent calculation of the centroid position will be subject to error. The situation is illustrated in Fig. 3.4 where the peak is assumed to have a simple isosceles triangular shape.

The time between samples is t^* and the time between the peak apex and the sample is t_1. Samples are taken from a fixed voltage threshold, V_d, to suppress the sampling of noise. For weak peaks, this threshold is relatively high, so the number of digital samples taken on the peak can be very small. Therefore the estimated position of the centroid will vary significantly with t_1. Halliday calculated the errors that might be expected in a bad case for this simple peak shape. For the calculation, it was assumed that $V_0 = 0.1$ V and $V_d = 0.05$ V, that scan conditions were 10 s/decade at resolving power 10 000 (peak width 434 μs) and that the sampling rate was 12.5 kHz, corresponding to two or three samples per peak depending on the value of t_1. Figure 3.5 shows the discrepancy between the estimated position of the centroid, cal-

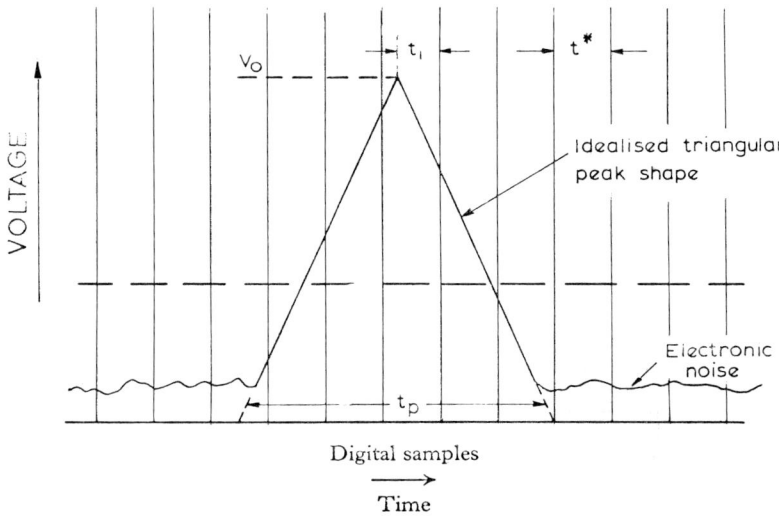

Fig. 3.4. Schematic representation of digitization of mass spectral peak. For details see text. (From ref. 18.)

culated from the two or three available digital samples and the actual apex of the triangle. The maximum possible error obtained (actually with three samples) is 5·4 p.p.m. and the standard deviation on a large number of individual peaks is about 3 p.p.m.

Halliday has termed this effect the "ratchet effect" for obvious reasons. It is evident that this error would be greatly reduced by the use of a faster digitization rate as well as by higher values of V_0 or a lower value of V_d.

Experiments carried out by Green et al. (19) and referred to by Halliday simulated a change of digitization rate (without any other changes) by first taking all digital samples for centroid calculations, then every alternate sample (half digitization rate) and then every fourth sample (quarter digitization rate). All the mass measurements thus found were subtracted from those found by using every digital sample. The data had been obtained using a scan rate of 84 s/decade in mass at a resolving power of 11 000 and a digitization rate of 10 kHz, equivalent to about 33 samples between the 5% points of every peak (provided it had sufficient intensity to be above threshold). As expected, it was found that the largest additional errors were introduced into the measurements on the smallest peaks, but even reducing the nominal number of samples per peak to eight added an error of only 2 p.p.m. standard deviation on the smallest peaks, in good agreement with the theoretical analysis. On larger peaks the error introduced was about 0·5 p.p.m. By putting 16 samples in a peak, the additional errors on small and large peaks were reduced to 1 p.p.m. and 0·3 p.p.m. respectively.

Bowen et al. (20) have reported that the use of more than about 20 samples across a peak gave no better accuracy in calculation of the centroid. Burlingame et al. (21) have more recently reported results using a scan rate of 8 s/decade at resolving power 10 000 and digitization rates of 50 and 100 kHz,

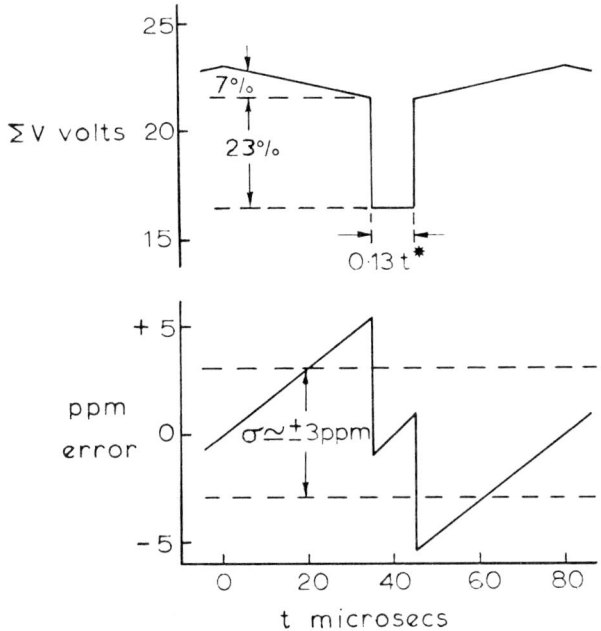

Fig. 3.5. Variation in estimated peak area, ΣV, and error in estimated peak position. Δ peak, V_0, 0·10 V, V_d 0·05 V. $\delta = 10 000$, $t_{10} = 10$ s, $t^* = 80$ μs, $t^* \equiv 18·4$ p.p.m. (12·5 kss^{-1}), $N = 3$ or 2 conversions. (From ref. 18.)

equivalent to about 18 and 36 samples per peak respectively. In these cases, peaks covering 100%–2% relative intensity gave standard deviations of 2·93 and 2·96 p.p.m. respectively. These experiments confirm that 20 digital samples per peak are more than sufficient to achieve good mass measurements and that 10 samples per peak are often enough.

3.4.5 Effect of Scan Rate and Resolving Power on Mass Measurement Accuracy

Finally, the effect of two instrumental factors, viz. scan time and resolving power, on accuracy of mass measurement, has to be considered. The effect of change of scan rate is to change the peak width and therefore the number of ions (N) to be introduced in eqn 3.11 proportionately to the scan time. As the resolving power is decreased, more ions will be required to establish a peak position with a given degree of certainty, as indicated by eqn 3.11.

If, in fact, sensitivity is inversely proportional to resolving power, it follows that $N \propto 1/R^2$ since not only the sensitivity but also the peak width decreases in inverse proportion to the resolving power. Substituting this relationship, we see that in this case σ is independent of resolving power for a given sample flow (57). Obviously, it would be advantageous if mass measurements could be carried out at low resolving power, because, whilst maintaining mass measurement accuracy, the use of a lower resolving power improves the limits of detection. In addition, the strain on data acquisition is relieved since the greater peak width allows lower sampling rates.

The main reason for the use of high resolving power in accurate mass measurement is to separate sample peaks from admixed reference peaks at the same nominal mass to ensure accurate centroid determination. In a double beam MS, this problem is overcome by using two separate sources, one for sample and one for reference. The beams from these two sources pass through common magnetic and electrostatic analyser sections and are then diverted again to separate amplification and recording systems. In this way, mass measurement at low resolving power and high sensitivity is possible (22). Naturally, if the sample spectrum itself contains unresolved multiplets at the resolving power employed, then these particular mass measurements will be in error. However, this is not such a common occurrence and, in any case, the parent ion, whose mass measurement will lead to the composition of the molecule, is usually a singlet in the spectrum of a pure compound. The process of calibration in double beam mass measurement is effected by introducing PFK into both sources, and taking a set of scans to determine the shift between the beams. During sample analysis, PFK is maintained in the reference beam, but is replaced by sample in the other beam. The sample is mass measured by interpolation between the reference beam peaks by the normal method previously described, but all mass measurements are then corrected by the shift appropriate to that mass given by the calibration. Distinction between sample and reference peaks is now trivial, unlike the normal single beam method (cf. p. 54).

An alternative approach to low resolution mass measurement, using a single beam instrument, has been described by Haddon and Lukens (23). In this method, after preliminary external calibration with PFK, the sample is run, under identical scanning conditions, admixed with a suitable internal calibrant, e.g. methylene iodide (CH_2I_2) or iodoform (CHI_3), that gives very few peaks, well resolved from most possible sample peaks.

A number of applications of mass measurement at low resolving power have been made, especially under fast scanning conditions, such as are required in capillary GCMS work and a reference to this type of work using double beam mass spectrometry is available (24). Without the ability to mass measure at low resolving power, a combination of fast scanning and high resolving power would impose impossible burdens on instrumental sensitivity and on digitization rates. The application of double beam mass measurement to chemical ionization spectra may also be noted.

3.4.6 Improvement of Mass Measurement Accuracy

The more accurate the mass measurement of any peak, the fewer the number of chemical compositions that have to be considered for that peak. Thus, the achievement of the highest possible accuracy is paramount in mass measurement work. The controlling factor in mass measurement accuracy is usually ion statistics (principally in the sample peak) and on this basis suggestions for improvement of accuracy have been made. Burlingame *et al.* (25) have suggested the simple expedient of averaging repetitive scans. They have, in fact, shown that in their case, averaging the results of n repetitive scans of the sample gives the expected overall improvement of $n^{\frac{1}{2}}$ in accuracy, thus demonstrating that the errors seen in single scans are mainly statistical and not systematic. Systematic errors would not show any improvement on averaging. In addition to an improvement in mass measurement accuracy, averaging offers a similar improvement in intensity measurement.

Another method for improvement of mass measurement accuracy has been suggested by McLafferty *et al.* (26). In their proposal, referred to as signal enhancement in real time (SERT), the time between peaks in a magnetically scanned high resolution spectrum is employed to effect successive re-scans of the first peak. This is done by making stepwise increments in the accelerating voltage and the electrostatic analyser (ESA) voltage in the same ratio after the peak has passed the collector so that, each time, the continuing magnet scan sweeps the peak over the collector again. Digital data is collected for each scan and this is averaged together to give a centroid figure for the peak. From a knowledge of the spacing of peaks in the spectra, the data system can be used to calculate the voltage increments needed and control the accelerating voltage and ESA voltage levels by applying these increments at the correct time. Just as in scan averaging if n scans are taken for each peak, the accuracy of mass measurement should increase by $n^{\frac{1}{2}}$ if the errors are entirely statistical. In preliminary reports of this system, improvements in mass measurement accuracy were obtained, but not at the levels hoped for. The reduced degree of improvement was ascribed to the presence of periodic noise. A more detailed analysis of the sources of error would be of interest.

3.5 PEAK-MATCHING METHODS

The classical technique of mass measurement with magnet sector machines is "peak-matching" (27). In this technique, a peak is generated by modulation of the ion beam across the collector slit by means of a small change in the ESA voltage or a small auxillary magnetic field. The accelerating voltage and ESA voltage of a double focusing instrument are also switched in the same ratio so that if a reference peak of known mass, M_R, is focused at the collector by an accelerating voltage, V_R, and the unknown mass, M, is focused at the collector by a voltage V, then:

$$M = \frac{M_R V_R}{V}. \tag{3.15}$$

The peak positions are usually compared visually by means of an oscilloscope display and the voltage ratio V/V_R adjusted until the two peaks are matched, i.e. coincident on the display.

As the improvements that can be made in peak-matching by use of a data system are much less dramatic than the acquisition of complete mass measurement data from a scan, little attention has so far been paid to this field. However, some workers in the field of accurate atomic mass determination have made use of a data system in peak-matching. For example, Duckworth et al. (28) have used off-line computer analysis to determine the matching condition. Their method involves switching to three slightly different voltages close to the expected voltage ratio for the unknown, and determining the centroid position of the unknown peak in each case as well as that of the reference peak using data accumulated by a signal averager. The matching condition, i.e. no shift of centroid between reference and unknown, is calculated from a straight line fit to this data. The whole procedure retains the improved signal to noise (S/N) ratio achieved by signal averaging, offers improved precision in a given operating time and further removes human judgement of the matching condition. Results were obtained for which the precision approached the statistical limit.

Hammar and coworkers (29) have reported the application of a data system based multiple peak monitoring (MPM) method to low resolution mass measurement of compounds eluted from a GC. In this method, the accelerating voltage is switched to any value between 3000 and 4000 V by adding a D/A converter controlled voltage to a basic accelerating voltage of 3000 V, giving a mass range of 30%. At the voltage level for both the reference and unknown masses, the data system also performs a narrow voltage sweep around the selected value. The sweep voltage consists of 51 increments of a selectable size and moves the ion beam over the collector slit. The ion signals are sampled during a variable integration time and are related to the increment number. Thus, a profile of the peak can be constructed and displayed and the mass determined by peak-matching from its centre position. The fastest sweep time (with an integration time of 1 ms) is 0·3 s which is stated to be fast enough even for capillary column work. At these speeds, a number of sweep cycles through the mass positions may be performed during the elution of a GC peak.

When peak matching is carried out at a low resolving power the choice of a reference compound that is separated from sample even at low resolving power, is necessary. Any unexpected, unresolved multiplets can be detected by visual examination of the recorded peak profiles. In practice, an initial run to determine an approximate mass using a wide sweep voltage followed by a more accurate determination using a narrow sweep proved necessary. Mass measurements showing a standard deviation of approx. 5 p.p.m. were obtained on the small parent ion of 1 µg of a basic compound eluted from the GC (29).

3.6 MASS MEASUREMENT FROM PHOTOPLATES

In organic mass spectrometry, the convenience and speed of operation of scanning MSs with electrical detection has resulted in the almost complete displacement of focal plane instruments from routine work. Focal plane detection has, however, remained in use in some areas that make maximum use of its merits, e.g. high resolution accurate mass measurement at ultimate sensitivity in organic analysis (30), the recording of transient spectra in field desorption analysis (31) and spark source inorganic trace analysis (Section 8.10).

For high resolution mass measurement work using photoplates, the sample is admixed with a reference compound just as when using a scanning MS. For conversion of times to masses, some point in the reference spectrum has to be known either from a previous calibration or from data typed in during the analysis. In theory, mass and distance on the photoplate should follow a simple quadratic relationship, i.e. $m^{\frac{1}{2}} \propto d$. This relationship is distorted by local effects, but holds sufficiently well over short distances for the position of a third reference peak to be predicted from those of two known, adjacent reference peaks. Once the next reference peak has been located, its true mass is assigned and all intervening sample peaks are mass measured assuming the same quadratic relationship.

Even this simple relationship is able to give results of high accuracy (32). Most other systems use a similar procedure, although in spark source work, where internal standards are limited to known matrix elements, more complex interpolation and extrapolation formulae are needed, particularly because of the greater distances between reference peaks (33). Again, in general, the accuracies in spark source mass measurement are lower because of the limited number of available reference peaks. Some of the most accurate results in organic analysis using a photoplate have been presented by Wadsworth and Tunnicliff (34) who used a least squares fit to the reference peaks to try to minimize fitting errors. Their results showed errors of 1–3 p.p.m. up to m/e 600.

3.7 THE RECOGNITION AND PROCESSING OF MULTIPLET DATA

A spectrum of a pure compound will sometimes contain two or more ions of different composition at the same nominal mass. If the instrumental resolving power used is insufficient, the peaks due to these ions will be incompletely separated and will be recorded as a multiplet. Some of the most common doublets including only carbon, hydrogen, nitrogen, oxygen and carbon-13 are listed in Table 3.4.

TABLE 3.4
Some common mass doublets

Doublet	Separation (m.m.u.)	Maximum mass at which the doublet is separated by 10 000 resolving power
$CH-^{13}C$	4·4692	44
N_2-CO	11·2335	112
CH_2-N	12·5754	125
NH_2-O	23·8089	238
CH_4-O	36·3843	363

When using a data system, the existence of a multiplet peak must be recognized before any treatment can begin. The simplest method of multiplet detection is inspection of the peak profile. Location of a valley that conforms to certain minimum width and depth criteria to distinguish it from inflections due to noise is usually taken to indicate the presence of a multiplet. The method is rapid and imposes no demands for storage of many digital samples. Thus, it can be executed on-line and used to flag suspected multiplets. However, the method is limited, especially as it will not detect poorly separated multiplets.

Storage of the complete profiles of all recorded peaks in a spectrum provides a basis for off-line recognition and treatment of multiplet data. This approach has been mainly applied to photoplate data where the full profiles are in any case available in a very compact form (34, 35, 37). However, other authors have used the data system to store complete digital profiles of all the peaks in a scan from an electrical recording system, although a large amount of storage is required in this method (15, 36). The method described by Desiderio (37) is one for off-line multiplet recognition and has been used with photoplate data. In this method, both the centroid and the apex (determined by curve fitting to the three highest points and taking the point where the first differential becomes zero) are computed for each peak. If these two values

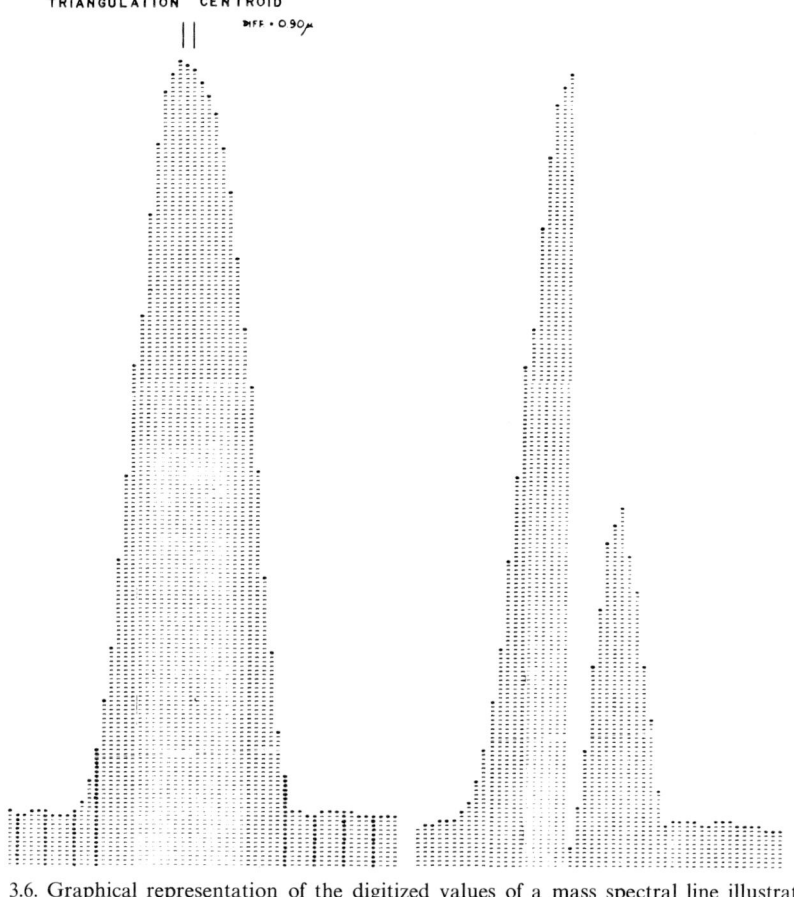

Fig. 3.6. Graphical representation of the digitized values of a mass spectral line illustrating "resolution" of multiplets. (From ref. 35b.)

differ by more than a threshold figure (generally 0·75 µm) the peak is assumed to be a multiplet (see Fig. 3.6). Another criterion for multiplet recognition is the ratio of peak height to width (35a).

Deconvolution of multiplet profiles off-line has usually been effected by an interative least squares fit method using either a Gaussian profile as representative of the recorded peak shape (34, 35a, 37, 38) or a known singlet profile (39). The application of this technique to the separation of a ^{13}C—CH doublet requiring a resolving power of 31 767 but recorded at a resolving power of 15 000 has been demonstrated (35a). A much less involved method of treating multiplet data was used by Biemann et al. (35b). The assumption made was that any multiplet was only a doublet. Then the steeper and therefore "purer" side of the peak profile was subtracted from the other side to leave the profile of the less abundant component of the doublet (Fig. 3.6). An effective resolution of 43 000 has been reported using this method (35b).

A method of compromise for use with an electrical recording system in which only the profiles of suspected multiplet peaks are stored by the data system has been suggested by Barber et al. (39). In this method, a peak is recognized as a multiplet, if its width is significantly greater than that of a singlet. The data system keeps a running check on the mean and standard deviation of the width at a fixed percentage peak height of successive peaks assumed to be singlets. For a peak whose width is significantly greater than these, the digital peak profile is stored for subsequent deconvolution. Doublets in which the smaller component was only 5% of the height of the larger one and which were separated by half a peak width were detected by this method and it was inferred that even with a separation of only one-third or one-quarter of a peak width the doublets would still have been detected. The method was most successful when peak widths could be determined at a suitably low percentage peak height, e.g. 5%.

3.8 THE ACQUISITION AND PROCESSING OF METASTABLE ION DATA

Most positive ions formed by electron impact will decompose, but with varying lifetimes. Those with half-lives of the order of 10^{-5} s are sufficiently stable to be accelerated out of the ionization chamber before decomposing. Many of these ions, however, decompose during their flight in the MS analyser tube and are called metastable ions. The product or daughter ions of these metastable decompositions appear in a mass spectrum as relatively small peaks which may be several mass units wide and are usually centered at non integral mass numbers (see Fig. 2.6). In double-focusing instruments, the metastable ions that decompose in the second field-free region—the region between the electrostatic and magnetic sectors (F_2 in Fig. 3.7)—give rise to the familiar metastable peaks normally seen in mass spectra with mass m^*, where:

$$m^* = m_2^2/m_1. \tag{3.16}$$

The acquisition and processing of metastable data from a normal low resolution scan raises a number of difficulties. Firstly, it is unusual for metastable peaks to reach 1% of the base peak intensity of the normal

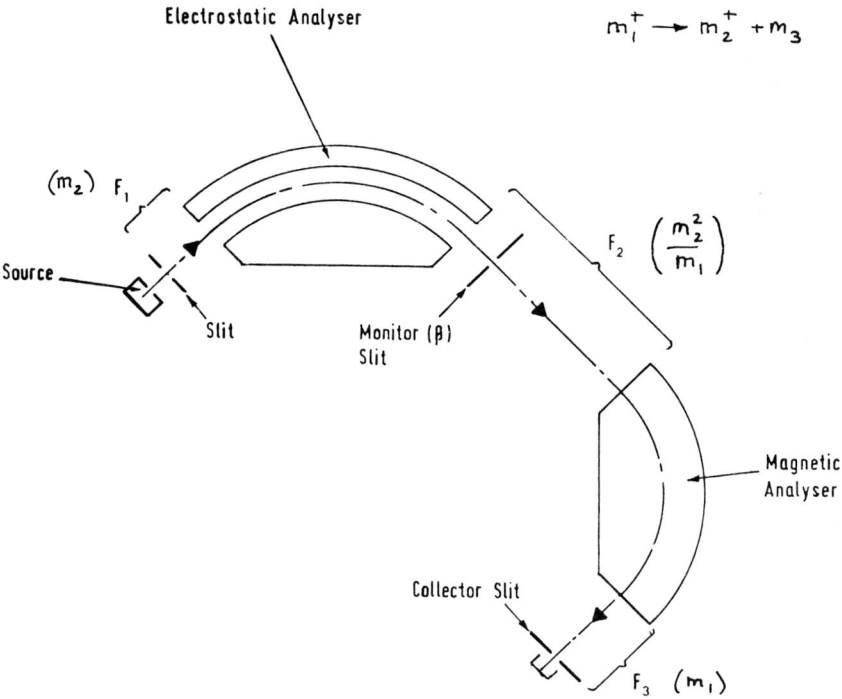

Fig. 3.7. Field-free regions in a double-focusing instrument of Nier–Johnson geometry. In region F_1, the ions formed will not pass through the electrostatic analyser unless V_{acc}/V_{esa} is increased by a factor m_1/m_2 (See text.)

spectrum and 0·1% is a more usual figure. Secondly, any procedure, such as the Hites–Biemann correction procedure that is designed to round off low resolution data to nominal and half masses will destroy any metastable mass data. Thirdly and perhaps most important, any program designed for the processing of metastable data must be able to separate out the effect of the unresolved and usually much more intense, normal peaks (Fig. 2.6).

A program for the separation of metastable data from normal scan data has been described by Duffield et al. (40). After passing a logarithmic amplifier, to ensure a more precise representation of low level signals, the spectrum was digitized and the normal peaks subtracted. A model of a normal peak for subtraction was obtained by averaging a number of the more abundant peaks in the spectrum of n-hexane run under identical conditions. Extensive testing was required to judge the success of the subtraction and this had to be carried out using a peak of closely defined intensity and of sufficient width to represent the tails of peaks. The residual metastable spectrum was then smoothed before analysis.

The authors recognized the difficulty in extracting metastable peaks present under intense normal peaks by this procedure. As an alternative, a procedure was developed in which the logarithmic spectrum could be displayed on a visual display unit and, using the skill of the operator to detect metastable ions, suggested metastables could then be analysed by the computer. Whichever recognition procedure was used, the analysis of this data displayed all the possible parent–daughter ion combinations for each metastable peak within a defined tolerance (cf. Section 7.6). However, it was still not possible to allocate, with certainty, the correct parent–daughter ion combination to each metastable peak.

Use of a defocusing method to observe pure metastable spectra overcomes both of the major disadvantages of working with normal spectra, i.e. interference from normal ions and uncertainty of identification of the corresponding transition. In addition, an enhancement of sensitivity results from the study of metastable ions formed in the first field-free region between the source and electrostatic sector (F_1 in Fig. 3.7) (43). As the ions formed in this region have a lower kinetic energy than normal ions they will not be transmitted by the electrostatic sector unless the accelerating voltage (V_{acc}) is increased relative to the electrostatic sector voltage (V_{esa}). Thus, by tuning in with the magnet to a normal ion at a lower value of V_{acc} and then scanning V_{acc} upwards whilst keeping V_{esa} constant, metastable ions corresponding to transitions from precursors of the normal ion will be observed. Linked scans of V_{acc} and V_{esa} and of V_{esa} and magnetic field have been used more recently to record metastable data (41). In these linked scans, peaks corresponding to the formation of daughter ions from the original ion are recorded.

Use has also been made of reversed geometry instruments in which the magnetic sector precedes the electrostatic sector. In this case, the ion of interest is selected using the magnet and then V_{esa} scanned to record metastable ions corresponding to the formation of daughter ions from the original ion. This technique is known as mass-analysed ion kinetic energy spectrometry (MIKES) (42).

Although not representing a fully automated method, the earliest attempts at automation of a defocusing method were made by Barber et al. (44). In these studies, normal daughter ions (m_2) were manually tuned in at the collector at a reduced accelerating voltage and this voltage then scanned up to the full value by means of a linear ramp generator. With a constant value of V_{esa}, precursor ions (m_1) were detected at a voltage $V = V_{acc} m_1/m_2$, where V_{acc} is the reduced accelerating voltage. Measurement of V was accomplished by slow digitization of the output and determination of the voltage corresponding to the centroid of each peak assuming the scan to be perfectly linear. A typical scan is shown in Fig. 3.8.

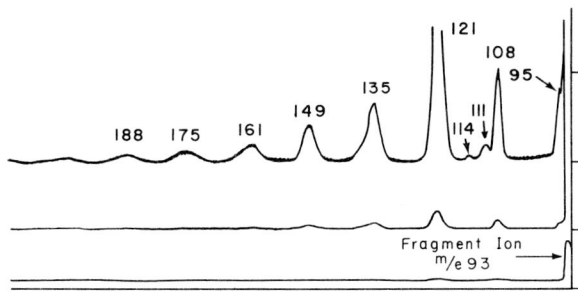

Fig. 3.8. Analogue record from metastable analysis by defocusing method. Analysis is of the precursor ions of m/e 93 in the spectrum of tomatidine. (From ref. 46.)

A report of data acquisition from accelerating voltage scans using the Carrick system has also appeared (45a). The same system was also applied to the acquisition of metastable data from normal high resolution scans (45). Coutant and McLafferty (46) have described a more complex system based on a suggestion by Major (47). If a normal metastable ion from the second field-free region ($m^* = m_2^2/m_1$) is focused at the collector and then the potential of the electrostatic sector is lowered, the corresponding metastable in the first field-free region may be detected. In this case, m_1 and m_2 may be uniquely determined from m^*, the mass of the normal metastable ion and $\rho = V'_{esa}/V_{esa}$ i.e. the reduced electrostatic sector voltage at which the defocused metastable ion is detected divided by the normal electrostatic sector voltage. Thus, $m^* = m_2^2/m_1$ and $m_2/m_1 = \rho$. In practice, the data system acquires spectra containing all possible metastable ions, as well as normal ions, by sweeping the electrostatic sector at a high rate compared to the sweep of the magnetic field. Thus each mass peak is covered by about 10 ESA scans. Data is digitized and values of mass and the ESA voltage are stored for later calculations. Although the system provides a complete coverage of metastable transitions, in practice a very slow scan (two hours for n-decane) is necessary if sufficient sensitivity is to be available for the detection of even a small fraction of these transitions.

Beynon and coworkers (48) have described a data acquisition system suitable for MIKES scans that employs a mechanical drive for the ESA voltage as well as a conventional digitizing system for the output data. It is stated that mechanical drive gives the most reproducible results, thus allowing signal averaging from successive scans. By signal averaging, a good S/N ratio could be observed even when working at high kinetic energy resolution. In this way, the determination of very small kinetic energy releases from the width of metastable peaks and the study of the fine structure of metastable peaks becomes possible.

3.9 THE ACQUISITION AND PROCESSING OF IONIZATION EFFICIENCY DATA

A simple modification to a conventional MS involving the fitting of a constant drive motor to the electron beam energy control allows the recording of ionization efficiency data (49). The ion for which the efficiency data is required is focused at the collector and the electron beam energy scanned downwards from a value near 14 eV at a scanning rate of about 10 eV/min. The collector signal is digitized at a rate of about 10 kHz or less. This data is averaged over approx. 100 points by a real time program and these averaged points temporarily stored. Further averaging reduces the data to about 200 points, typically at 0·03 V intervals. 10 such scans are taken and the data from these averaged and smoothed.

Automated measurement of ionization efficiency data has also been described using a quadrupole MS with a conventional electron impact ion source (50) and Gross and Wilkins (51) have described an ion cyclotron resonance spectrometer-computer system with particular reference to appearance potential measurements. The energy spread in the incident electron beam associated with conventional electron impact sources has necessitated the development of mathematical techniques for the deconvolution of ionization efficiency curves obtained in this way. A comparison of two such techniques, namely, the Fourier transform method (52) and the electron distribution difference method (53) found the Fourier transform method to be considerably superior for the recovery of fine structure and for the determination of accurate ionization and appearance potentials (54). A simpler method for the determination of ionization and appearance potentials from this type of data has been used more recently by Johnstone and McMaster (56). Alternatively, automated versions of experimental methods using a quasi-monoenergetic electron beam have been reported (55).

REFERENCES

1. Plattner J. R. and Markey S. P. (1971). *Org. Mass Spectrom.* **5**, 463.
2. Hites R. A. and Biemann K. (1967) *Analyt. Chem.* **39**, 965.
3. Binks R., Cleaver R. L., Littler J. S. and MacMillan J. (1971). *Chem. in Britain* **7**, 8.
4. Grayson M. A. and Conrads R. J. (1970). *Analyt. Chem.* **42**, 456.
5. Jansson P. Å., Melkersson S., Ryhage R. and Wilkström S. (1969). *Ark. Kemi* **31**, 565.
6. Griffith D. A. (1972). *In* "Biochemical Applications of Mass Spectrometry" (Ed. Waller G. R.), p. 58. John Wiley and Sons, Chichester.
7. (a) Sweeley C. C., Ray B. D., Krichevsky M. I. and Holland J. F. (1972). *In* "Biochemical Applications of Mass Spectrometry" (Ed. Waller G. R.), p. 52. John Wiley and Sons, Chichester.
   (b) Sweeley C. C., Ray B. D., Wood W. I., Holland J. F. and Krichevsky M. I. (1970). *Analyt. Chem.* **42**, 1505.
8. Houseman J. and Hafner F. W. (1971). *J. Phys.* (E), **4**, 46.
9. Barber M., Chapman J. R., Green B. N., Merren T. O. and Riddoch A. (1970). 18th Annual Conference on Mass Spectrometry and Allied Topics, San Francisco. Paper L8, p. B299.
10. (a) Johnson M. W., Gordon B. W. and Self R. (1973). *Lab. Pract.* **22**, 267.
    (b) Smith D. H., Olsen R. W., Walls F. C. and Burlingame A. L. (1971). *Analyt. Chem.* **43**, 1796.
    (c) Burlingame A. L. (1968). *In* "Advances in Mass Spectrometry" (Ed. Kendrick E.), Vol. 4, Institute of Petroleum, London. p. 77.
11. (a) Holmes W. F., Holland W. H. and Parker J. A. (1971). *Analyt. Chem.* **43**, 1806.
    (b) Hedfjäll B., Jansson P. A., Marde Y., Ryhage R. and Wilkström S. (1969). *J. Phys.* (E) **2**, 1031.
    (c) Schwarzenbach R. and Clerc J. T. (1973). *Org. Mass Spectrom.* **7**, 1215.
12. McMurray W. J. (1972). *In* "Biochemical Applications of Mass Spectrometry" (Ed. Waller G. R.), p. 70. John Wiley and Sons, Chichester.
13. Bowen H. C., Clayton E., Shields D. J. and Stanier H. M. (1968). "Advances in Mass Spectrometry" (Ed. Kendrick E.), Vol. 4, p. 257. Institute of Petroleum, London.
14. (a) Ahmed Z., Deinet W., Gold W. and Wölfel U. (1971). *Siemens-Z.* **45**, 585.
    (b) Klimowski R. J., Venkataraghavan R. and McLafferty F. W. (1970). *Org. Mass Spectrom.* **4**, 17.
15. Christie W. H., Smith D. H. and McKown H. S. (1973–4). *Chem. Instrum.* **5**, 43.
16. Burlingame A. L., Chang J. J. and Holland P. T. (1972). 20th Annual Conference on Mass Spectrometry and Allied Topics, Dallas. Paper O7, p. 287.
17. Smith D. H., Olsen R. W. and Burlingame A. L. (1968). 16th Annual Conference on Mass Spectrometry and Allied Topics, Pittsburgh. Paper 37, p. 101.
18. Halliday J. S. (1968). *In* "Advances in Mass Spectrometry" (Ed. Kendrick E.), Vol. 4, p. 239. Institute of Petroleum, London.
19. McMurray W. J., Lipsky S. R. and Green B. N. (1968). *In* "Advances in Mass Spectrometry" (Ed. Kendrick E.), Vol. 4, p. 77. Institute of Petroleum, London.
20. Bowen H. C., Chenevix-Trench T., Drackley S. D., Faust R. C. and Saunders R. A. (1967). *J. scient. Instrum.* **44**, 343.

21. Kimble B. J., Cox R. E., McPherron R. V., Olsen R. W., Roitman E., Walls F. C. and Burlingame A. L. (1974). 22nd Annual Conference on Mass Spectrometry and Allied Topics, Philadelphia. Paper P10, p. 315.
22. Aspinal M. L., Chapman J. R., Compson K. R., Hazelby D. and Riddoch A. (1973). 21st Annual Conference on Mass Spectrometry and Allied Topics, San Francisco. Paper T13, p. 471.
23. Haddon W. F. and Lukens H. C. (1974). 22nd Annual Conference on Mass Spectrometry and Allied Topics, Philadelphia. Paper U2, p. 436.
24. Aspinal M. L., Compson K. R., Dowman A. A., Elliott R. M. and Hazelby D. (1975). 23rd Annual Conference on Mass Spectrometry and Allied Topics, Houston. (1975). Paper C2, p. 73.
25. Burlingame A. L., Smith D. H., Merren T. O. and Olsen R. W. (1970). *In* "Computers in Analytical Chemistry", *Prog. Analyt. Chem.* Vol. 4, p. 17. Plenum Press, New York.
26. McLafferty F. W., Michnowicz J. A., Venkataraghavan R., Rogerson P. and Giessner B. G. (1972). *Analyt. Chem.* **44**, 2282.
27. Nier A. O. (1957). *In* "Nuclear Masses and their Determination" (Ed. H. Hintenberger). Pergamon Press, Oxford.
28. Meredith J. O., Southon F. C. G., Barber R. C., Williams P. and Duckworth H. E. (1972–3). *Int. J. Mass Spectrom. Ion Phys.* **10**, 359.
29. Hammar C-G., Pettersson G. and Carpenter P. T. (1974). *Biomed. Mass Spectrom.* **1**, 397.
30. Biemann K. (1972). *In* "Biochemical Applications of Mass Spectrometry" (Ed. Waller G. R.), p. 96. John Wiley and Sons, Chichester.
31. Schulten H-R. (1974). *Biomed. Mass Spectrom.* **1**, 223.
32. Desiderio D. and Biemann K. (1964). 12th Annual Conference on Mass Spectrometry and Allied Topics, Montreal. Paper 69, p. 433.
33. Frisch M. A. and Reuter W. (1973). *Analyt. Chem.* **45**, 1889.
34. Tunnicliff D. D. and Wadsworth P. A. (1968). *Analyt. Chem.* **40**, 1826.
35. (a) Venkatarhagavan R., McLafferty F. W. and Amy J. K. (1967). *Analyt. Chem.* **39**, 178.
    (b) Cone C., Fennessey P., Hites R. A., Mancuso N. and Biemann K. (1967). 15th Annual Conference on Mass Spectrometry and Allied Topics, Denver. Paper 39, p. 114.
36. Burlingame A. L., Olsen R. W. and McPherron R. (1974). *In* "Advances in Mass Spectrometry" (Ed. West A. R.), Vol. 6, p. 1053. Applied Science Publishers, Barking.
37. Desiderio D. M. (1971). *In* "Mass Spectrometry, Techniques and Applications" (Ed. Milne G. W. A.), p. 11. John Wiley and Sons, Chichester.
38. Von Meerwall E. D. and Gawlik M. D. (1974). *Comput. Phys. Communs.* **7**, 115.
39. Barber M. and Green B. N. (1968). 16th Annual Conference on Mass Spectrometry and Allied Topics, Pittsburgh. Paper 35, p. 91.
40. Duffield A. M., Reynolds W. E., Anderson D. A., Stillman R. A. and Carroll C. E. (1971). 19th Annual Conference on Mass Spectrometry and Allied Topics, Atlanta. Paper D4, p. 63.
41. (a) Weston A. T., Jennings K. R., Evans S. and Elliott R. M. (1976). *Int. J. Mass Spectrom. Ion Phys.* **20**, 317.
    (b) Bruins A. P., Jennings K. R. and Evans S. *Int. J. Mass. Spectrom. Ion Phys.* (submitted).
    (c) Boyd R. K. and Beynon J. H. *Org. Mass Spectrom.* (in press).

42. Beynon J. H., Cooks R. G., Amy J. W., Baitinger W. E. and Ridley T. (1973). *Analyt. Chem.* **45**, 1023A.
43. Barber M. and Elliott R. M. (1964). 12th Annual Conference on Mass Spectrometry and Allied Topics, Montreal. Paper 22, p. 150.
44. Barber, M., Green B. N., Wolstenholme W. A. and Jennings K. R. (1968). *In* "Advances in Mass Spectrometry" (Ed. Kendrick E.), Vol. 4, p. 89. Institute of Petroleum, London.
45. (a) Carrick A. (1972). 20th Annual Conference on Mass Spectrometry and Allied Topics, Dallas. Paper T1, p. 407.
    (b) Carrick A. and Paisley H. M. (1974). *Org. Mass Spectrom.* **8**, 229.
46. Coutant J. E. and McLafferty F. W. (1972). *Int. J. Mass Spectrom. Ion Phys.* **8**, 323.
47. Major H. W. (1968). Docket No. D-1185, U.S. Patent Office, Washington D.C.
48. Terwilliger D. T., Beynon J. H. and Cooks R. G. (1974). *Int. J. Mass Spectrom. Ion Phys.* **14**, 15.
49. Johnstone R. A. W., Mellon F. A. and Ward S. D. (1970). *Int. J. Mass Spectrom. Ion Phys.* **5**, 241.
50. Dromey R. G., Morrison J. D. and Traegar J. C. (1971). *Int. J. Mass Spectrom. Ion Phys.* **6**, 57.
51. Gross M. L. and Wilkins C. L. (1971). *Analyt. Chem.* **43**, 1624.
52. Morrison J. D. (1963). *J. Chem. Phys.* **39**, 200.
53. Winters R. E., Collins J. H. and Courchene W. L. (1966). *J. Chem. Phys.* **45**, 1931.
54. Giessner B. G. and Meisels G. G. (1970). 18th Annual Conference on Mass Spectrometry and Allied Topics, San Francisco. Paper S3, p. B431.
55. Lifshitz C., Agam J., Weinberg A., Kantor D., Shainok U. and Peres M. (1973). *Int. J. Mass Spectrom. Ion Phys.* **11**, 243.
56. Johnstone R. A. W. and McMaster B. N. (1973). *J.C.S. Chem. Comm.* 730.

4
Data Reduction

4.1 INTRODUCTION

In this chapter, we deal with the processes of data reduction. Data reduction is defined to include the reporting of recorded spectra in tabular or graphical form, the reporting of time dependent data such as total ion current (TIC) chromatograms and the calculation and presentation of elemental composition data from accurate mass measurements. Data reduction also includes the location of "important" spectra from the wealth of data recorded in the repetitive scanning mode and the identification and resolution of mixture spectra.

4.2 NOMINAL MASS DATA—SCANNING METHODS

In the analysis of a pure sample, it is a simple matter to record one scan when, for example, the TIC reaches a suitable level. Even when a somewhat impure sample, e.g. the preparation of a trimethylsilyl ether, is fractionally evaporated from the solids probe, it is relatively easy to judge when to record a satisfactory spectrum of the major component by observation of an oscilloscope display of repeated scans. In these straightforward cases, where the number of components of interest is very small—often one—the operator is able to follow simple rules and recognize spectral patterns to provide a highly efficient filtering of the potential data output of the mass spectrometer.

However, where the number of expected components exceeds one or two, for example, in most analyses by combined gas chromatography—mass spectrometry (GCMS) or in the fractionation of more complex mixtures from the solids probe, the operator is no longer so efficient in deciding when and when not to record scans. In these situations, use of the continuous scanning technique (1) is preferred. In this technique, the scans are not initiated at times chosen by the operator, but are repetitive, usually fast, scans (e.g. a scan of 1 s duration repeated every 2 s) initiated by the data system for the duration of sample analysis. A further advantage of this technique that applies to the analysis of any sample is that the complete record of the sample analysis which becomes available allows a retrospective search for any previously unsuspected sample component or for unsuspected charges in the recorded spectrum of a labile compound.

4.3 NOMINAL MASS DATA—PRESENTATION

The presentation of a nominal mass spectrum once recorded, takes two standard forms, a tabulation of mass and intensity data (Fig. 4.1) and the graphical form or plot of mass against intensity (Fig. 4.2). Whichever method

```
NO. PEAKS:     82
BASE/NREF INT:   11214./  11214.
TIC:       53790.
MASS RANGE:    0  -   138
RETN TIME/MISC:   40:28/   17/   0/   5/   0
```

NOMINAL MASS	ABSOLUTE INTENSITY	% INT. BASE
138.	170.	1.5
137.	181.	1.6
136.	2454.	21.9
122.	188.	1.7
121.	2233.	19.9
120.	117.	1.0
108.	346.	3.1
107.	1753.	15.6
106.	115.	1.0
105.	761.	6.8
95.	802.	7.2
94.	2069.	18.5
93.	6632.	59.1
92.	1897.	16.9
91.	1161.	10.4
82.	163.	1.5
81.	1252.	11.2
80.	963.	8.6
79.	2351.	21.0
78.	301.	2.7
77.	1206.	10.8
69.	664.	5.9
68.	11214.	100.0
67.	3888.	34.7
66.	720.	6.4
65.	366.	3.3
55.	505.	4.5
53.	1688.	15.1
52.	218.	1.9
51.	451.	4.0

Fig. 4.1. Partial tabulation of mass and intensity data from spectrum of limonene shown in Fig. 4.2.

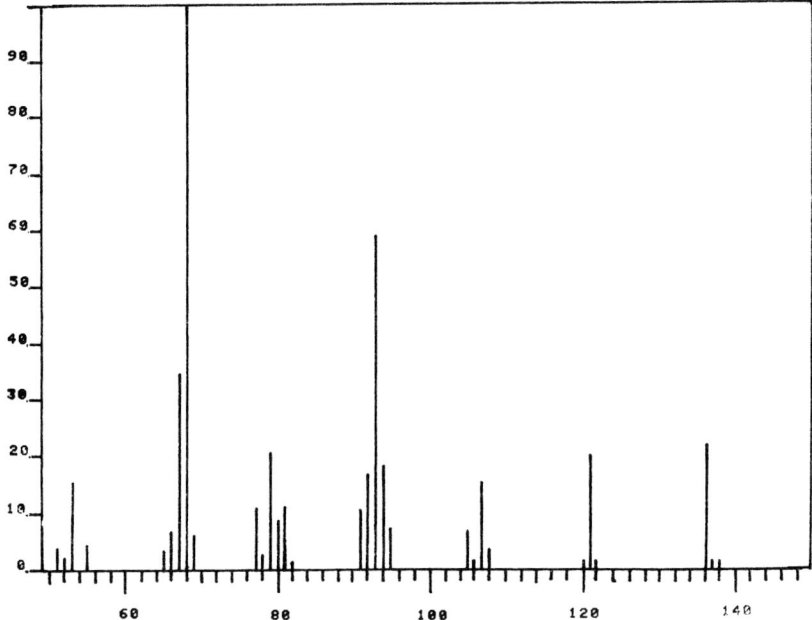

Fig. 4.2. Plotted spectrum of limonene corresponding to Fig. 4.1.

is used for presentation, the spectrum must first have been placed on the same scale as other appropriate spectra by normalization. The most common normalization procedure is to make the largest peak in each spectrum equal to 100 intensity units (normalization to the base peak). This method is not entirely satisfactory because it depends to a large extent on the correct measurement of the one peak most likely to be overloaded, i.e. the base peak. Despite this possible drawback, this is the method generally chosen. Other procedures are possible, e.g. normalization so that the intensity of any specified peak equals 100 or normalization so that the sum of all the peak intensities equals one. The latter procedure has the advantage that normalization is no longer entirely dependent on correct measurement of base peak intensity.

4.4 REDUCTION AND CLEAN-UP OF REPETITIVE SCAN NOMINAL MASS DATA—MIXTURE ANALYSIS

When using the continuous scanning technique, the immediate problem is to decide which of the many recorded spectra are the most relevant ones to the analytical problem in hand. This is particularly true when the technique is applied to the GCMS analysis of complex mixtures. In GCMS applications, a summary of the analysis is usually provided in the form of a TIC chromatogram. This is generated by summing absolute intensities of all peaks recorded in each spectrum. The sum for each spectrum is then normalized to the maximum value and these sums plotted v. time. As the scans are taken at regular and frequent intervals in time, the plot provides a reasonable sample intensity profile (Fig. 4.3) equivalent to a normal gas chromatogram.

Restrictions may be placed on the masses used for plotting the TIC chromatogram. For example, it may exclude masses below m/e 45 to exclude background contributions due to air and water in the MS or it may exclude notably large background peaks (e.g. m/e 207 due to bleed from silicone GC

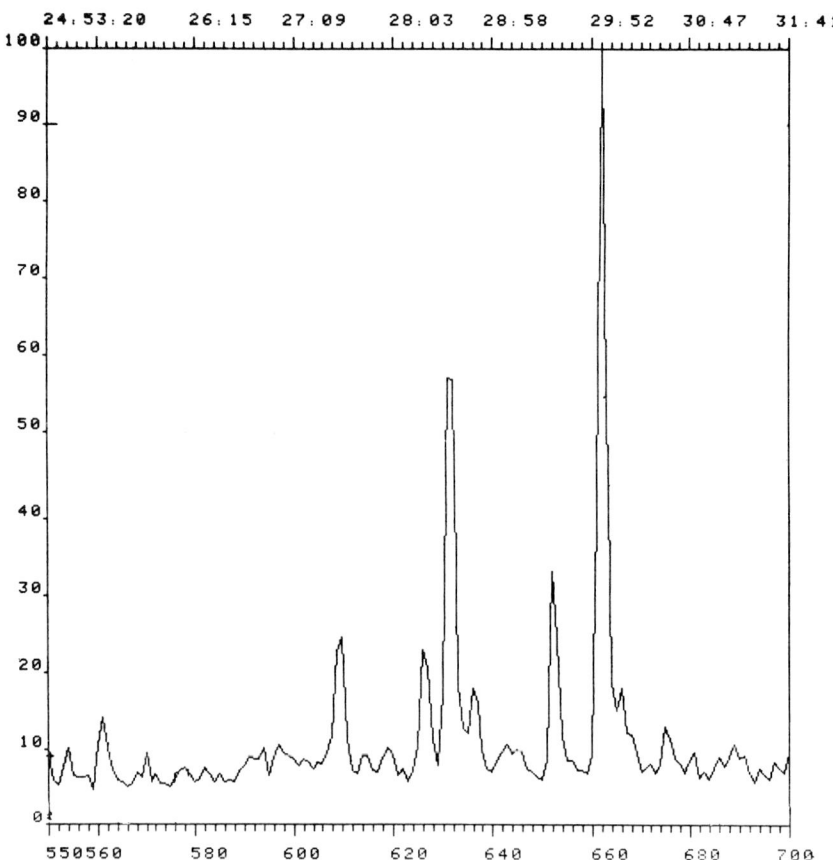

Fig. 4.3. Part of total ion current chromatogram from glass capillary column analysis of tobacco smoke (Courtesy Kratos-AEI Scientific Instruments).

stationary phases). If a general analysis of an unknown mixture by GCMS is required, in the absence of any more specific directions it would be usual to inspect spectra from peak maxima and inflections as displayed on the TIC chromatogram. A method designed to automatically choose "important" spectra using on-line TIC data and so reduce the number of spectra that had to be stored was described by Henneberg et al. in 1972 (2). A logic diagram

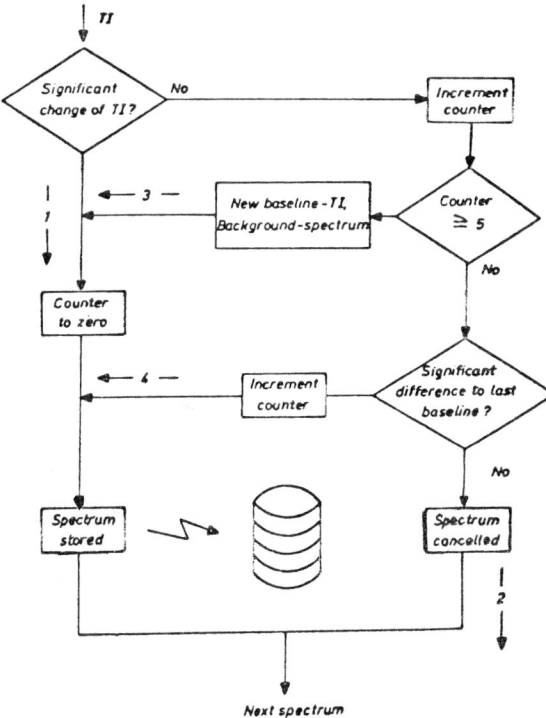

Fig. 4.4. Logic diagram of algorithm for the selection of important spectra. (From ref. 2.)

of their algorithm is shown in Fig. 4.4 and the 4 possible paths through the diagram are numbered 1–4. These paths may be described as follows:

Path 1. If the TIC of a spectrum differs significantly from that of the preceding one, the spectrum is stored. A significant difference is equivalent to the TIC of a spectrum whose base peak has a signal to noise ratio of 100:1.

Path 2. If TIC does not change significantly, increment the counter. So long as the counter is below five and the TIC value does not differ significantly from the recently defined baseline TIC value, the spectrum is cancelled.

Path 3. If the counter reaches five, the spectrum is stored. This system provides a satisfactory record of changes in the baseline and also provides a spectrum for background correction of subsequent Path 1 spectra.

Path 4. A TIC value may not be significantly different from the preceding one when, for example, successive spectra are taken on each side of the narrow GC peak. In this case, however, the TIC value differs from the baseline value and the spectrum has to be stored. The same conditions may occur on the flat part of a tailing GC peak. In this case, after Path 4 has been run through a few times, the increased counter value will cause a branch into Path 3, whereupon a new baseline TIC value will be set and spectra will no longer be stored.

The algorithm provides a convenient, automatic method of reducing the amount of data collected. In addition, the number of the last spectrum reached by Path 3 is stored in a data file with each of the following spectra. This spectrum is then automatically identified without further input as the appropriate spectrum to use for background subtraction of the following spectra. The algorithm provides convenience of operation and will produce savings on memory of perhaps 50% in some cases. However, real time rejection of some spectra can prevent a subsequent detailed inspection of the chromatogram for the presence of unresolved components. Thus recording of all spectra may be preferred particularly with the continuing reduction in the unit cost of bulk memory.

In performing subtraction of background spectra, the background and sample spectra are usually normalized to put them on an equivalent intensity basis. Normalization is usually carried out to the base peak of the background spectrum, disregarding such peaks as m/e 28 which may be overloaded. An alternative procedure omits normalization and involves only the direct subtraction of absolute intensities. Whilst these subtraction procedures are used to correct for background such as column bleed or air, water and perhaps hydrocarbons present in the MS, they are also applicable to the resolution of mixed spectra, i.e. spectra of co-eluting or co-evaporating samples. An excellent example of both applications of subtraction is given by Reimendal and Sjövall (3) in dealing with a complex spectrum from the GCMS analysis of steroids in human urine. Subtraction of background from the recorded spectrum leaves the corrected spectrum (Fig. 4.5). If the previous, background corrected scan is subtracted from (a) spectrum (b) results. If the succeeding background corrected scan is subtracted from (a), spectrum (c) results. Thus, it can be seen that spectrum (a) is the sum of the two other spectra, and, in fact, examination of scan (b) shows that it too is probably the sum of two spectra of compounds of mol. wt 728 and 683. Spectrum (c) appears to be the spectrum of a single compound of mol. wt 595.

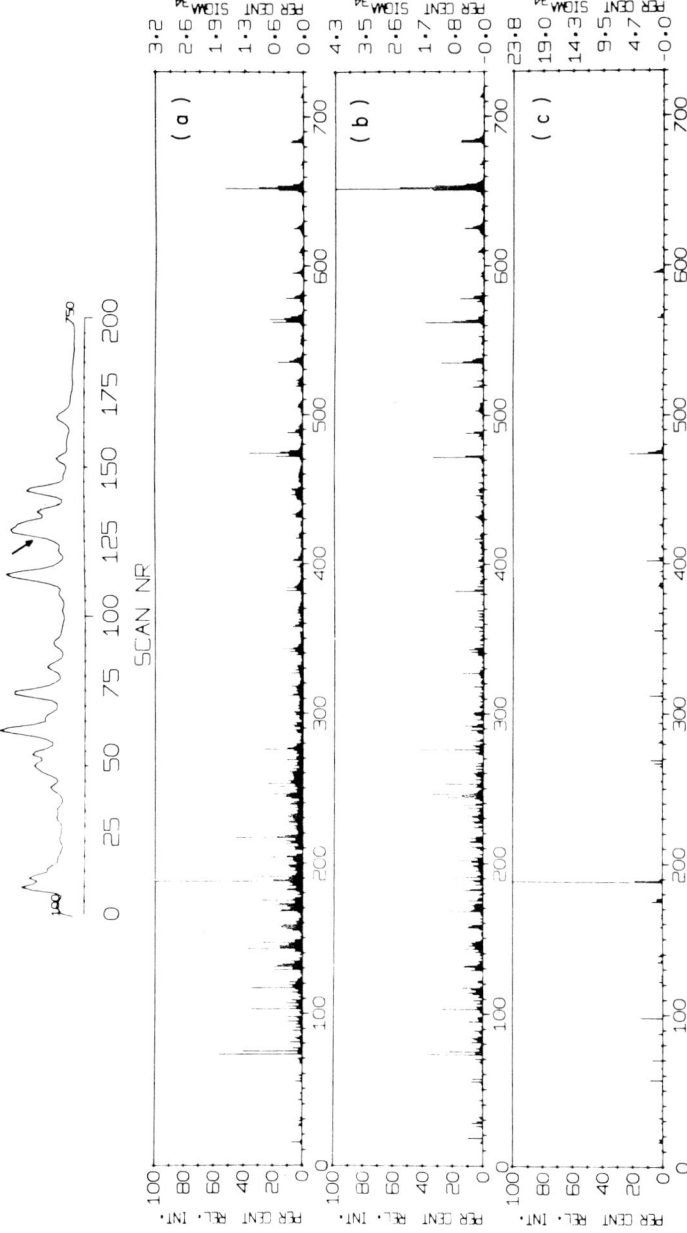

Fig. 4.5. Spectra obtained at scan 126 in the analysis of MO–TMS derivatives of steroids from urine of a healthy male. Top trace: Partial ion current chromatogram, m/e 100–750. Spectrum (a): Background corrected spectrum 126. Spectrum (b): Difference spectrum between background corrected spectra 126 and 125. Spectrum (c): Difference spectrum between background corrected spectra 126 and 127. (Reprinted with permission from ref. 3. Copyright by the American Chemical Society.)

A successful background subtraction or resolution of a mixed spectrum is usually a prerequisite for spectral identification or interpretation whether this is performed manually or by a computer-based technique. Unfortunately, it is rarely possible to carry out a perfect background subtraction by the simple techniques so far described because of the fluctuation of background and sample intensities, so that the procedures described by Biller and Biemann (4) (p. 84) Reimendal and Sjövall (3) (p. 89) and particularly by Dromey et al. (5) (p. 89) are important in offering viable alternative techniques.

In the case where each successive spectrum has been stored, other methods of data reduction are available. One such method is the use of the mass chromatogram introduced by Hites and Biemann (6). The mass chromatogram is a display v. time, of the intensity recorded at a specific mass value. For example, the parent ion of hexa-2,5-dione (m/e 114) is reasonably intense (8–9%) and thus the display of m/e 114 as a mass chromatogram should show a strong response from the presence of hexa-2,5-dione, but hopefully a much diminished response from other components indicated by the TIC chromatogram (Fig. 4.6). Thus, the mass chromatogram presents the MS output as that of a specific and highly sensitive detector. In general, the use of this technique demands some prior knowledge of the compound or compound type sought, although in the system employed by Biemann, mass chromatograms are computed for all masses after the analysis and recorded, together with the TIC chromatogram trace, on microfilm (7). Thus, whilst the use of mass chromatograms is in many cases a very valuable approach, in the general case when all possible mass chromatograms are plotted, the operator is faced with an overwhelming amount of data for analysis.

Mass chromatograms other than those of single masses have also been used. The homologous series chromatogram has been described by Biemann and coworkers (7b). Homologous series chromatograms plot the sum of the intensities of ions differing by 14 a.m.u. against scan number. Another variant described by Eichelberger et al. (8) is a plot of the sum of the intensities of a subset of masses characteristic of a range of compounds. Use of this technique with a subset of masses (m/e 190, 224, 260, 294, 330, 362 and 394) characteristic of polychlorinated biphenyls (PCBs) allowed the detection of PCBs from monochlorobiphenyl to hexachlorobiphenyl in environmental samples (Fig. 4.7).

More recently, Biller and Biemann (4) have described a more automatic and general analysis of mass chromatogram data. In this method, a peak profile analysis is performed by the data system on the mass chromatogram for each mass of the entire mass range covered. After completion of this process, the resulting array contains, for each scan number, only those m/e

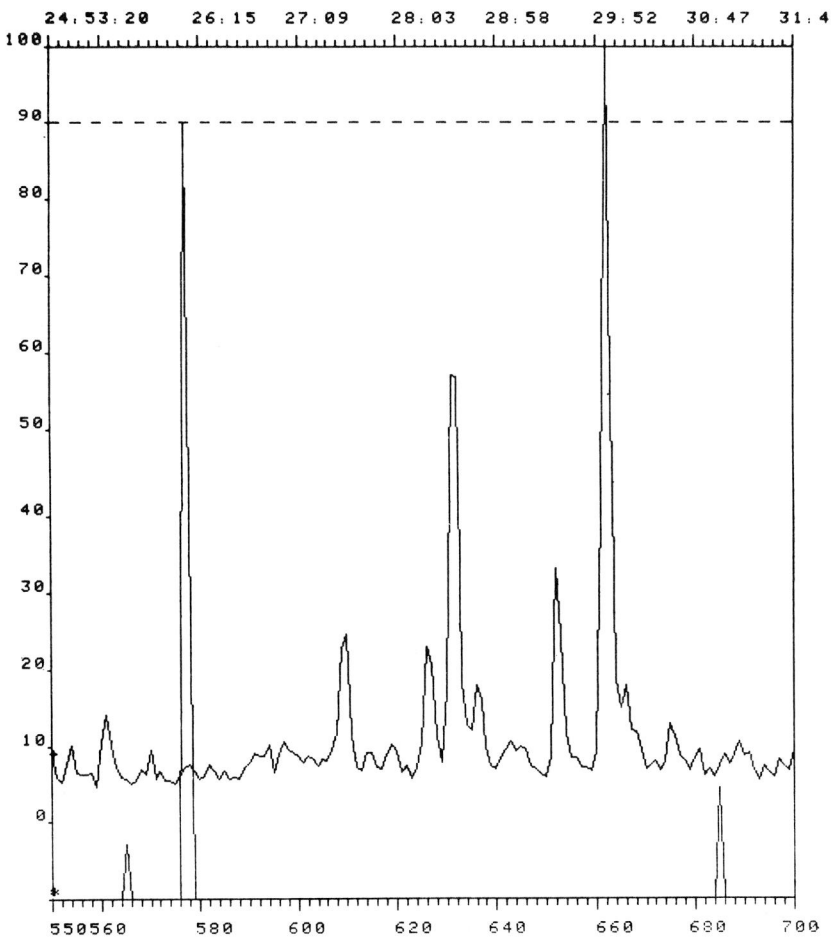

Fig. 4.6. Mass chromatogram of m/e 114 vs. total ion current, showing tentative identification of hexa-2, 5-dione in tobacco smoke (scan 577) (compare Fig. 4.3). (Courtesy Kratos-AEI Scientific Instruments).

values which maximize in that scan, along with their absolute intensities. Each of these spectra is referred to as a reconstructed mass spectrum. It is sometimes possible for different ions from the spectrum of a single compound to maximize in adjacent scans, especially when sample concentrations eluting from the GC undergo rapid change. For this reason, in the final output, each spectrum is the sum of the reconstructed mass spectra of scan n and scan $n + 1$, so that the eventual resolution of the system may be somewhat diminished. Finally, the absolute intensities of all ions maximizing in

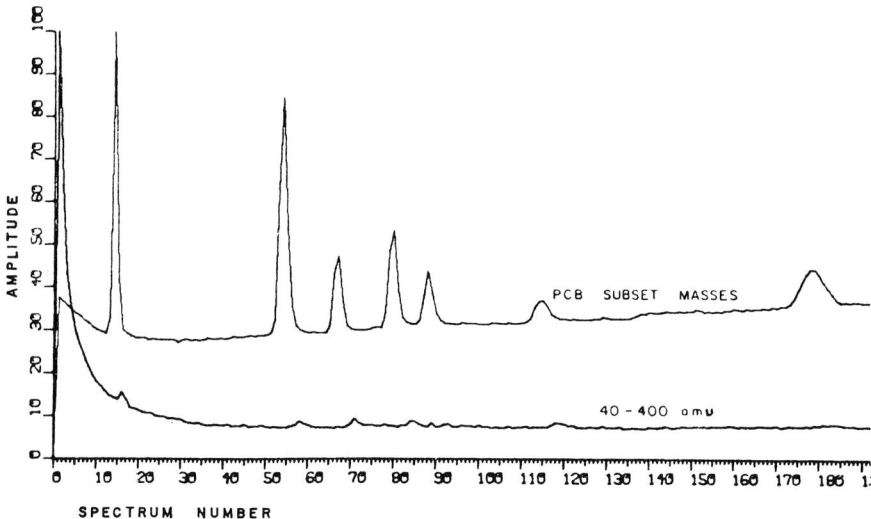

Fig. 4.7. Ion abundance chromatograms of a mixture of 5 ng of each of 7 pure PCB isomers. Lower chromatogram monitors masses 40–400; upper chromatogram monitors the PCB subset masses. (Reprinted with permission from ref. 8. Copyright by the American Chemical Society.)

a given scan are summed and plotted against scan index number to yield a set of data that is analogous to a TIC chromatogram plot and for which Biller and Biemann propose the term mass resolved gas chromatogram.

The data in Fig. 4.8 shows the conventional TIC plot and the mass resolved gas chromatogram from the analysis of an extract of a gastric lavage (4). Comparison of the two traces shows a dramatic increase in the number of components visible. From this kind of analysis, by inspection of spectra corresponding to the maxima and inflections in the mass resolved gas chromatogram, it is possible to approach an exhaustive analysis of complex mixtures by GCMS. Naturally, this technique is applicable to other situations, e.g. fractionation of a sample from a solids probe, where repetitive scanning is used. Biller and Biemann also suggest that these reconstructed mass spectra rather than the original spectra, may with advantage, be used in their library search technique (9) (p. 128).

An example of reconstructed mass spectra, taken from the work of Biller and Biemann (4), is given in Fig. 4.9. Reconstructed spectrum 112 represents the masses maximizing in scans 111 and 112. As can be seen, the lack of change in sample composition results in a blank spectrum. Reconstructed spectrum 113 indicates the beginnings of change in the original scan 113. Reconstructed spectrum 114 shows a reconstructed spectrum of docosane

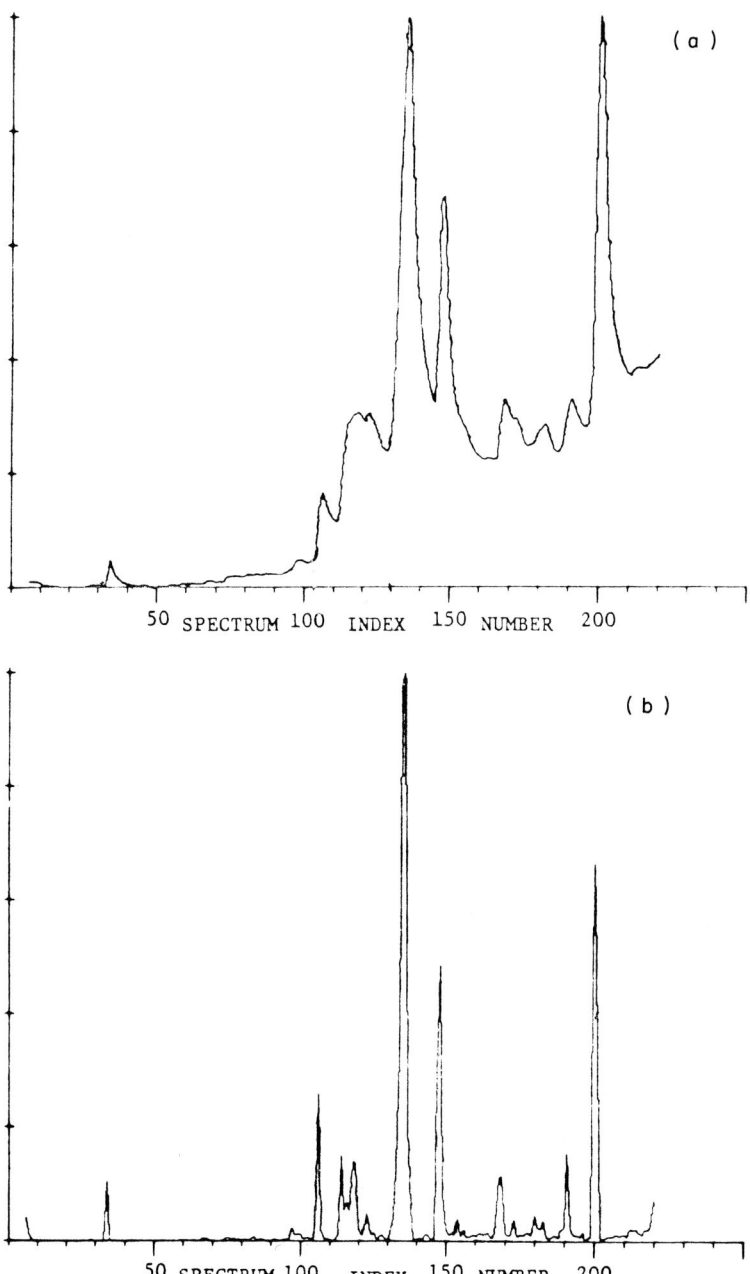

Fig. 4.8. (a) Conventional gas chromatogram (as total ion current plot). (b) The corresponding mass resolved gas chromatogram. (Reprinted from ref. 4, by courtesy of Marcel Dekker, Inc.)

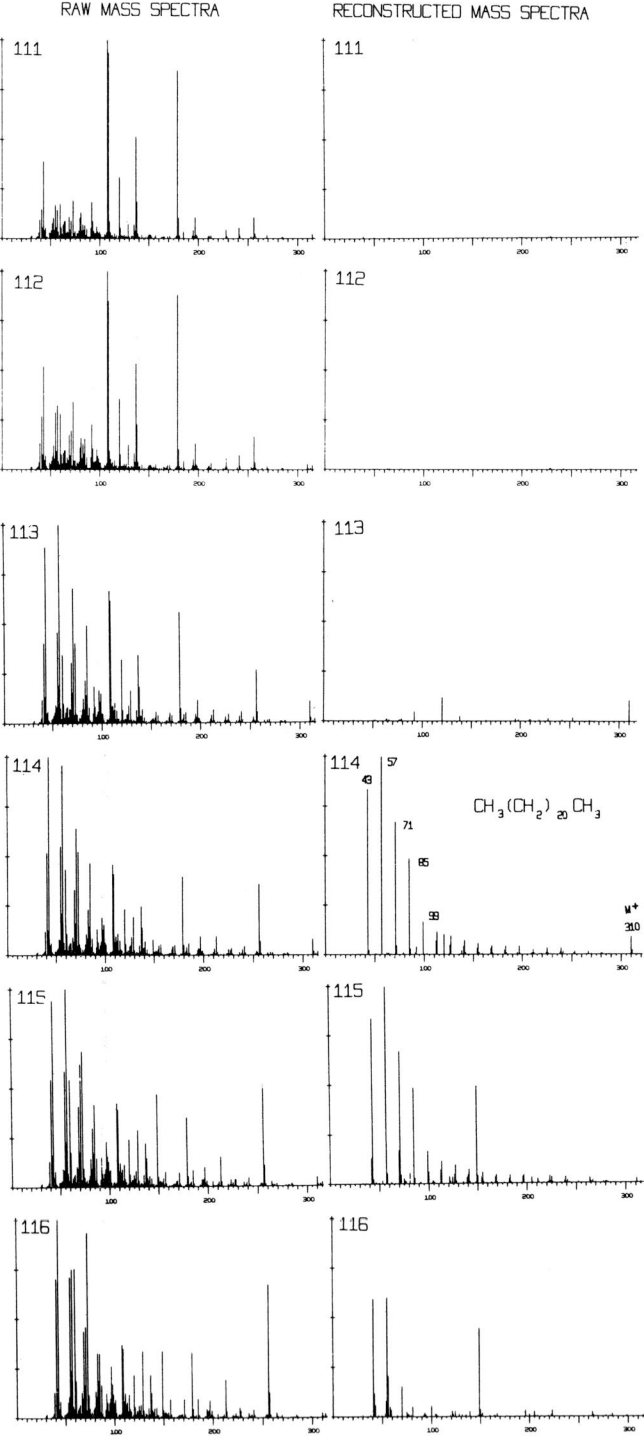

which began to emerge in scan 113 and maximized in scan 114. It may be noted that peaks at m/e 149 (phthalate) and m/e 256 (palmitic acid) which appear in the original scan 114 are excluded from the reconstructed spectrum because they belong to components whose intensity is still increasing at this stage. Although some of the minor peaks of the hydrocarbon spectrum (e.g. m/e 41 + 14n series) are lost on reconstruction, nevertheless the characteristic features of this type of spectrum are clearly retained.

A similar approach involving the location of mass resolved maxima had previously been described by Reimendal and Sjövall (3). Again, the changes with time of the intensities of all fragment ions are followed and although in fact a complex algorithm is used, essentially, when a given number of fragment ion intensities maximize in a single scan with above a certain intensity, that scan is deemed to be a scan of interest. The program also has provision to ignore major background ions. As Reimendal and Sjövall point out, this system can detect overlapping compounds provided that the peak maxima differ by at least one scan and their mass spectra are different. This cannot be achieved using the TIC trace and, indeed, the method is much more sensitive than is the use of the TIC. The peak locating program does fail with overlapping compounds that give similar spectra, e.g. mixtures of isomers or mixtures of steroids differing only in their degree of unsaturation. The authors continue their examination of located scans using an empirical program especially designed to elicit structural information in their field of application, viz. steroid analysis.

A very sophisticated method for the location of components and the subsequent extraction of representative spectra free from background peaks and peaks from unresolved components, particularly as applied to GCMS analysis, has been described by Dromey *et al.* (5) as part of the Stanford project. Component maxima are located by computing histograms of the numbers of maximizing masses in each scan and of the total maximizing ion intensity in these scans. The actual maximum position is determined by a parabolic least squares interpolation about the top five points in this data and the time calculated to one-third of a scan time to resolve very close neighbours. Calculation of background corrected intensities is achieved by the assumption of two models, a background that varies linearly with time over small regions and a model elutant peak taken from the experimental data itself. Using these models, a least squares estimate of the corrected intensity is made for every mass which meets a number of conditions, principally that it should maximize near to the interpolated maximum. Information about a component that has just been detected is stored

---

Fig. 4.9. Raw and reconstructed mass spectra from the analysis illustrated in Fig. 4.8. (Reprinted from ref. 4, by courtesy of Marcel Dekker, Inc.)

temporarily and if the histograms then show a further component within a certain number of scan widths, an algorithm for doublet resolution is initiated. In this case a least squares fit to two model peaks on a linear background is used. Using this procedure, the resolution of components within one and a half to two scan intervals has been demonstrated. Thus, characterisation of the data on a finer scale than can be achieved by the method of Biller and Biemann is possible.

Working as an off-line procedure, the Stanford program will analyse 600 spectra in approx. 8 min and reduce them to a set of "clean" component spectra. Examples given in the paper include resolution of a component that comprises only 4% of the total recorded ion current at its maximum (Fig. 4.10). Another important feature of this "Cleanup" program is that it is also able to correct automatically for peak saturation by use of the model described (Fig. 4.11).

Work described by Haber (10) used a model peak to extract useful information from GCMS runs. In this case a Gaussian profile was used as a matched filter for mass chromatogram data. Not only was the technique useful as a means of data reduction but was also a good method for estimating the location of very poorly resolved GC components. The technique was applied to the resolution of a mixture of acetylcholine-d_0 and acetylcholine-d_6 showing less than 1% difference in retention time when analysed by GCMS.

Another technique which may be applied to mixture analysis is factor analysis, also referred to as principal component analysis or eigenvector analysis (cf. Section 6.7). Factor analysis is a multivariate statistical technique which may be used to determine the number of independent variables contributing to measurements, in this case the number of compounds contributing to a mixture spectrum. The method was first applied by Davies *et al.* (11) to demonstrate the presence of two components in unresolved mixtures of hexane and heptane analysed by GCMS, where only masses common to both spectra were considered to make the problem more complex. Using data from successive scans, a co-variance or correlation matrix relating the intensity at each mass position to that at all other mass positions in the scan is formed. Analysis of this matrix yields independent factors that are linear combinations of these correlations. Successive factors account for the greatest possible residual variance and these factors are extracted until the residual variance reaches a level corresponding to experimental error. The number of factors extracted represents the number of independent components of the mixture. It should be emphasized that this application of factor analysis makes no assumptions regarding GC peak shape or the nature of the components or their spectra. A more detailed description of the application of factor analysis to mixture analysis, with examples, has been given by Ritter et al. (12) and the method has also been used by Halket and Reid (13). A fully worked example of factor analysis may be found in the appendix to a paper by Simmonds (14) and a discussion of methodology in the paper by Rozett and Petersen (15).

Rosenthal and Bursey (16) have reported a method that permits enhancement of the information present in scans when searching for the presence of particular compounds eluted from a gas chromatograph. The method utilizes the representation of a low resolution mass spectrum as a multidimensional vector, the intensity of each peak being equivalent to a distance along one of the mutually perpendicular axes in the vector space (cf. pp. 150 and 157). For an unknown spectrum, $U = (U_1, U_2, \ldots, U_n)$, and the standard spectrum, $S = (S_1, S_2, \ldots, S_n)$, the angle between the two vector representations is given by eqn 4.1:

$$\theta = \cos^{-1} \frac{\sum_{i=1}^{n} U_i S_i}{\left(\sum_{i=1}^{n} U_i U_i \sum_{i=1}^{n} S_i S_i\right)^{\frac{1}{2}}} \tag{4.1}$$

and the projection of U on S by eqn 4.2:

$$U_{proj} = |U| \cos \theta. \tag{4.2}$$

Plots of θ and U_{proj} v. spectrum number were made and inspected for minima and maxima respectively. Whilst the value of θ can be correlated with identity with the component sought ($\theta = 0$ for perfect identity), the projection of the unknown spectrum on the component sought gives a quantitative estimate of that component. Rosenthal and Bursey observed a sharp discrimination against components of the mixture which were not sought, together with a 25:1 enhancement of signal to noise (S/N) ratio using this method.

4.5 ELEMENTAL COMPOSITIONS FROM NOMINAL MASS DATA—CALCULATION OF MONOISOTOPIC SPECTRA

The presence and sometimes the numbers of elements with more pronounced isotope patterns may be inferred from an examination of intensity distributions in nominal mass spectra. Chlorine, bromine, sulphur and boron are amongst the many elements whose presence may be recognized in this way. A number of authors have reported programs which calculate nominal mass numbers and relative intensities of each of the lines in the mass cluster

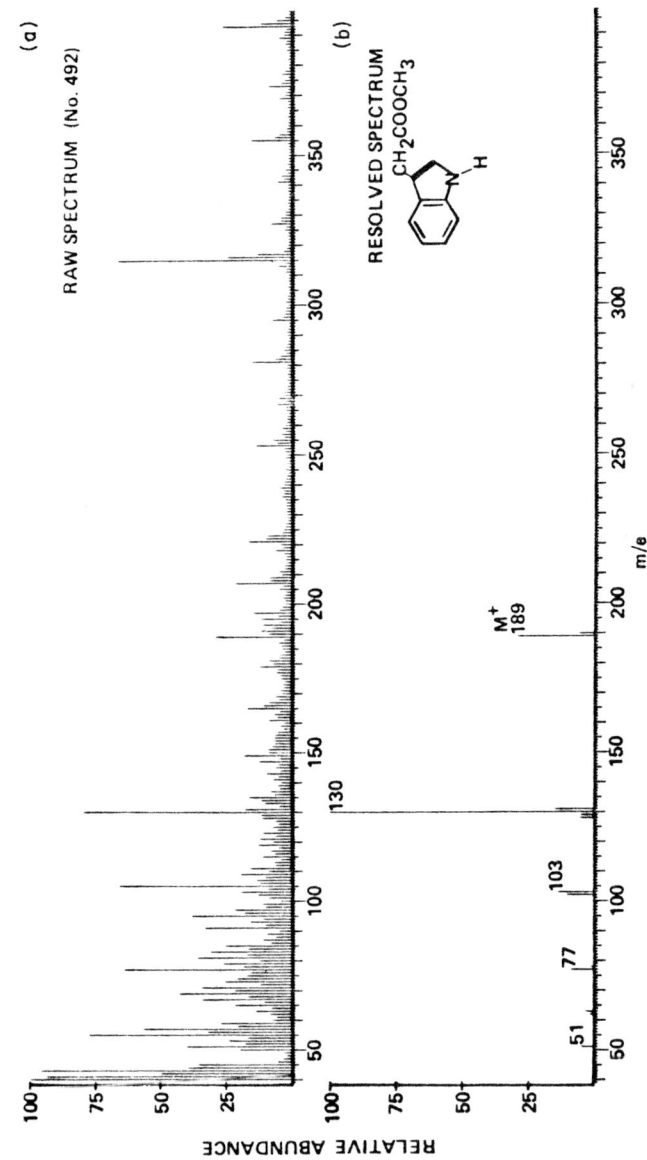

Fig. 4.10. (a) Spectrum of indole acetic acid 3-methyl ester from a GCMS analysis of human urine before processing. (b) Resolved spectrum of indole acetic acid 3-methyl ester. (Reprinted with permission from ref. 5. Copyright by the American Chemical Society.)

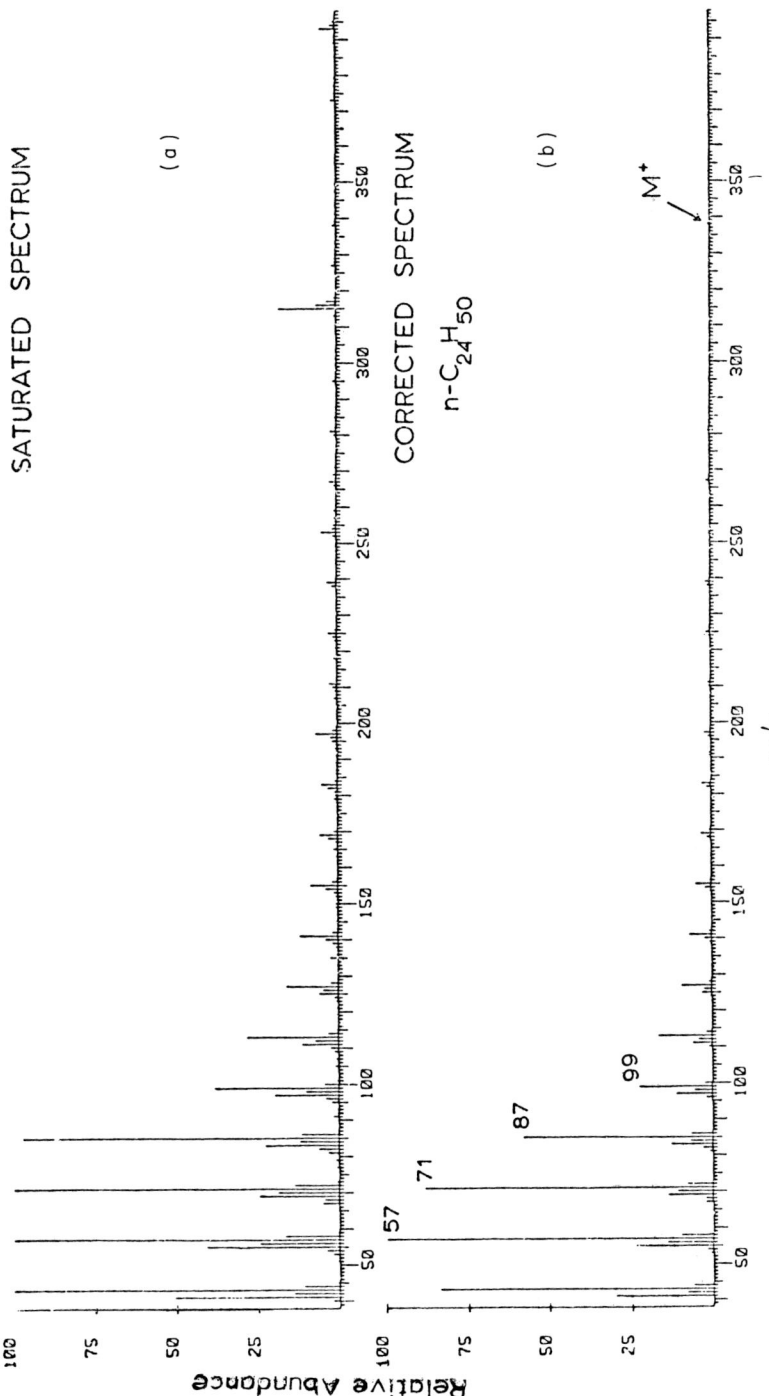

Fig. 4.11. (a) Saturated unprocessed mass spectrum for tetracosane. (b) Spectrum of tetracosane after processing and correction for saturation. (Reprinted with permission from ref. 5. Copyright by the American Chemical Society.)

corresponding to an ion containing one or more polyisotopic elements (17–21). An example of the output from one such program is shown in Fig. 4.12.

```
C    H    O    N    S    P    F    CL   BR   J    B    SN                                       KODE  DRUCK
24   9    **   6    **   **   **   4    **¡  **   1    **   **   **   **   CL  B ( CN CN C6H3 CL )3    -0    NUR

LINIENZAHL        GESAMT =   49      INTENS.MINDEST.  0.0010   PROZENT DES BASISPEAKS =  13      SUM =  0.9999997

MASSE   INTENSITAET      INTENSITAET      MASSE   DIAGRAMM
        SUMME = 1        BASISPEAK=100

531     0.047726         17.0546          531     ---------
532     0.209590         74.8953          532     ------------------------------------
533     0.120956         43.2228          533     ---------------------
534     0.279844         100.0000         534     --------------------------------------------------
535     0.107395         38.3769          535     -------------------
536     0.142707         50.9954          536     -------------------------
537     0.044677         15.9651          537     --------
538     0.033752         12.0611          538     ------
539     0.009034         3.2284           539     --
540     0.003458         1.2357           540     -
541     0.000753         0.2691           541
542     0.000097         0.0347           542
543     0.000008         0.0030           543
```

Fig. 4.12. Output of program ISOTOP showing isotope peak intensities for the formula ClB (CNCNC$_6$H$_3$Cl)$_3^+$. (From ref. 19).

More sophisticated programs have been written that, given a full polyisotopic nominal mass spectrum and a suggested molecular composition will check the recorded intensities for goodness of fit using known isotopic abundancies (22–23). The method described by Crawford (23) begins by enumerating all possible monoisotopic ions formed from elements contained in the suggested composition that could contribute to the observed spectrum. The method of McLaughlin and Rozett (22) requires the formulae or choice of formulae for the ions present as input. The isotopic mass spectrum of each ion is then generated and a least squares fit to the observed spectrum attempted using these ions. The goodness of fit obtained by this procedure can be used to determine whether the suggested composition is a probable one or not or whether there are impurities present in the sample. For example, McLaughlin et al. (24) showed by this means that a spectrum reported as that of $B_{20}H_{26}$ results from a mixture of $C_9B_{10}H_{28}$ and probably $C_9B_9H_{29}$. It should perhaps be emphasized that the application of these programs in this way has been limited to the study of spectra of compounds containing elements having abundant isotopic peaks such as boron, bromine and chlorine. These programs may also be used to derive monoisotopic spectra, i.e. spectra in which only the contributions due entirely to principal isotopes are retained, from recorded spectra. Monoisotopic spectra represent a considerable simplification of the original data whilst retaining all the information on fragmentation. Another application of these programs is to calculate isotopic abundances from recorded spectra (25). Other related programs have also been reported (26–27).

An attempt has been made by Evans and Jurinski (28) to devise a program that will compute an elemental analysis for an ion cluster based on the nominal mass and relative intensity of the cluster ions and considering the elements C, H, N, Cl, Br, F, S, Si, O and P. Because the natural abundance of the nitrogen isotope, ^{15}N, is at the level of experimental error, the presence or absence of this element is decided by the operator. The interactive routine also allows the operator to correct any errors the program may be thought to have made.

A program due to Semenov et al. (29) develops a table of empirical formulae satisfying a mol. wt determined from a mass spectrum and also recommends the smallest set of elements which must be determined experimentally (e.g. by micro-analysis) to unambiguously establish the true empirical formula. Any prior knowledge of the elemental composition is also accepted by the program in reducing the number of possible formulae.

4.6 REDUCTION OF ACCURATE MASS DATA— CONVERSION TO ELEMENTAL COMPOSITIONS

Many of the data reduction techniques described so far as being relevant to nominal mass data, e.g. normalization, subtraction and clean-up techniques are also relevant to accurate mass data. However, in practice, with the exception of normalization, these techniques have been much less used for treatment of this type of data, principally because accurate mass measurement has been too little employed in the GCMS field where these data reduction techniques are most often used. The major data reduction step applied to accurate mass data is its conversion to a list of possible elemental compositions corresponding to each measured mass. Programs have been written to calculate elemental compositions consistent with a measured mass (30, 31) and a complete program listing is given by Shrader (30). Tables of accurate masses corresponding to elemental compositions have been published (32, 33) and various tables allowing manual assignation of a composition to a measured mass have also been published (34, 35).

The additional input required for a program to calculate elemental compositions corresponding to a measured mass is a list of the type and maximum number of atoms to be used in generating the compositions together with a mass measurement tolerance. It has been customary to define this tolerance in parts per million (p.p.m.), as the theoretical mass measurement accuracy for a given peak size expressed in this way is not mass dependent unlike a definition in milli-mass units (mmu). However, from the chemical viewpoint, the error is more realistically expressed in mmu as, for example, the separation of the CH_4—O doublet expressed in this way is independent of mass, unlike its expression in p.p.m. For a given p.p.m. error and list of elements to be considered, the number of possible compositions increases rapidly with increasing mass. Thus it is often necessary to suggest an upper limit to the error in mmu as well, to keep the output within reasonable limits. For example, the tolerance could be 10 p.p.m. or 3 mmu, whichever is first exceeded.

An elemental composition routine must be able to list and calculate the mass of every possible combination of atoms within the limits suggested by the operator. However, a number of these possible combinations may be excluded on valency considerations. A simple formula may be used to calculate the double bond equivalent, D (i.e. number of double bonds and rings) for any suggested composition, viz:

$$D = 1 + 0{\cdot}5 \sum_{i=1}^{i_{max}} N_i(V_i - 2) \quad (4.3)$$

where N_i is the total number of atoms of valency V_i. If it is accepted that an example of the most saturated ion that can exist is a protonated saturated alcohol $(C_nH_{2n+3}O)^+$, then D must have a minimum value of $-0{\cdot}5$ i.e.:

$$\sum_{i=1}^{i_{max}} N_i(V_i - 2) \geqslant -3. \quad (4.4)$$

Equation 4.4 may be used to check compositions after calculation or it may be used to limit the calculations themselves. In the latter case, it leads to a formula (eqn 4.5) that defines a minimum value for the number of carbon atoms (N_C) that may be contained at any mass for a specified combination of atoms other than hydrogen. The formula is:

$$N_C \geqslant \frac{M_H(A - B)}{2M_H + M_C} \quad (4.5)$$

where M_H and M_C are the accurate masses of hydrogen and carbon respectively and A and B are given by:

$$A = \frac{M - \sum_{i=1}^{i_{max}} N_i M_i}{M_H} \qquad B = 3 - \sum_{i=1}^{i_{max}} N_i(V_i - 2).$$

Thus, for a mass M, which is the measured mass less the measurement tolerance, the program considers all possible combinations of specified heteroatoms other than carbon and hydrogen. In each combination, there are N_i atoms of type i having an accurate mass M_i and a valency V_i and which are not carbon or hydrogen. Equation 4.5 gives a minimum number of carbon atoms but there is also an obvious upper limit given by eqn 4.6:

$$N_C \leqslant \frac{M' - \sum_{i=1}^{i_{max}} N_i M_i}{M_C} \qquad (4.6)$$

where M' is the measured mass plus the measurement tolerance. Part of a typical atomic composition report is shown in Fig. 4.13. The output may be limited, as may nominal mass reports, by specification of a mass range and by specification of a minimum peak intensity for reporting. There is also a dramatic decrease in the size of output as the number and type of heteroatoms to be considered is decreased.

CALCULATED MASS	DEV	C12/13	H	O	N	MEASURED MASS	NO. PTS	REL INT
225.0659	-2.9	17/1	8	0	0	225.0629	6	1.8
225.0637	- .7	10/0	11	5	1			
207.0320	1.3	13/0	5	2	1	207.0333	4	.5
180.0741	- .4	9/1	11	3	0	180.0737	4	.4
179.0708	.6	10/0	11	3	0	179.0714	8	6.7
179.0734	-2.0	13/0	9	0	1			
179.0689	2.4	12/1	8	0	1			
167.0496	2.5	12/0	7	1	0	167.0521	4	.3
167.0537	-1.6	7/1	8	3	1			
166.0503	-2.2	8/0	8	3	1	166.0482	8	5.8
166.0458	2.3	7/1	7	3	1			
140.0302	- .5	5/1	5	3	1	140.0297	5	.4
139.0269	.4	6/0	5	3	1	139.0273	9	7.0

Fig. 4.13. Partial atomic composition output from a nitro and methoxyl substituted methyl ester of phenyl acetic acid. Correct identifications are underlined. Note that silicone material at m/e 207 may be recognized by the sudden change in fractional mass.

In addition the size of output naturally decreases as the mass measurement accuracy is increased (see Fig. 4.14). However, an increase in accuracy is often not possible so that an effort must be made to restrict the number and type of atoms to be considered. One approach to this problem involves the specification, as input, of submolecular groups such as phenyl (C_6H_5) carbonyl (CO) or methylene (CH_2) rather than atoms in an attempt to limit the amount of print out (39).

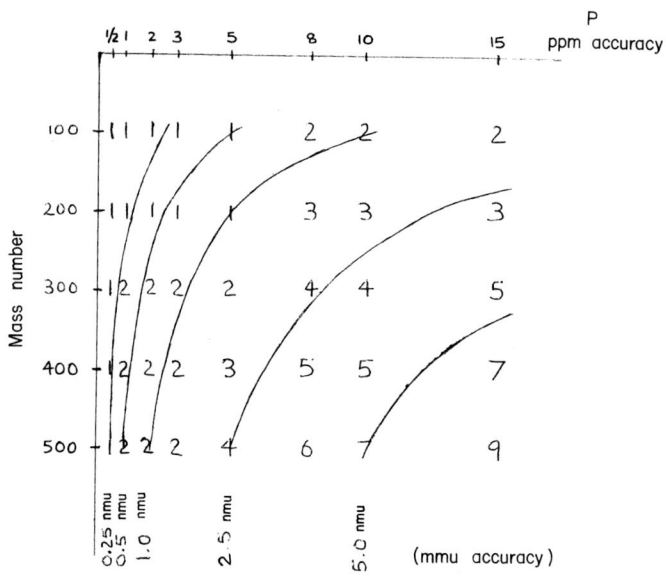

Fig. 4.14. Figure showing maximum number of possible compositions containing up to six oxygen and six nitrogen atoms for a molecule of given nominal mass whose exact mass has been measured with an accuracy of $\pm P$ p.p.m.

Besides background knowledge relevant to the composition of the sample, the presence or absence of the appropriate isotope peaks may be used as a preliminary check on whether to include certain elements amongst those to be considered. For example, the interactive system described by Biemann et al. (36) checks for the presence of pairs of peaks 1·998 ± 0·003 a.m.u. apart within rather coarse intensity tolerances before admitting Cl, Br, Si and S for consideration. Other checks are used to decide whether oxygen or nitrogen may be present (see p. 197).

The program described by Hilmer and Taylor (37) has been designed to make a more exhaustive analysis of an accurate mass spectrum using intensity data. Their procedure converts all isotopic formulae into a single chemically equivalent non-isotopic formula. A weighting factor is derived for each formula which expresses how many formulae at lower mass that particular formula contains (e.g. $H_4C_6NCl_3$ contains $H_3C_6NCl_2$ but not $H_3C_4N_2Cl$). Thus, the weighting factor should be highest for those formulae which correlate best with the rest of the spectrum at lower mass. Not only does the conversion of the recorded spectrum to the corresponding non-isotopic spectrum reduce the total output considerably, but also the weighting factor may be used to eliminate a large number of false formulae on the basis of poor correlation with the rest of the spectrum. It is normally found that the odd-electron species remaining with the largest weighting factor corresponds to the parent ion species.

Other, much simpler, methods for using intensity data to check or limit compositions derived from measured masses have been published. In an earlier paper, Hilmer and Taylor (38) suggest that any proposed formula containing isotopic atoms should only be accepted if all isotopic homologues of that formula which are more abundant can also be found in the spectrum. For example, $C^{37}Cl^+$ should not be accepted unless the more abundant $C^{35}Cl^+$ is also present. In addition, any isotopic combination with calculated relative abundance less than 2×10^{-4} is automatically rejected. The program described by Bell (21) derives support for suggested elemental compositions by confirming that the relative intensity values are correct. A useful program described by Varmuza and Krenmayr (19) calculates the accurate mass and the relative intensity for each isotopic component based on a given composition. The program also gives the multiplicity at each nominal mass and a resultant mass and intensity is given, assuming the components at that mass to be unresolved. A similar program has been reported by Carrick and Glockling (17).

REFERENCES

1. Hites R. A. and Biemann K. (1968). *Analyt. Chem.* **40**, 1217.
2. Henneberg D., Casper K. and Ziegler E. (1972). *Chromatographia* **5**, 209.
3. Reimendal R. and Sjövall J. B. (1973). *Analyt. Chem.* **45**, 1083.
4. Biller J. E. and Biemann K. (1974). *Analyt. Letts* **7**, 515.
5. Dromey R. G., Stefik M. J., Rindfleisch T. C. and Duffield A. M. (1976). *Analyt. Chem.* **48**, 1368.
6. Hites R. A. and Biemann K. (1970). *Analyt. Chem.* **42**, 855.
7. (a) Biemann K. (1972). *In* "Biochemical Application of Mass Spectrometry" (Ed. Waller G. R.), p. 96. John Wiley and Sons, Chichester.
   (b) Biller J. E., Hertz H. S. and Biemann K. (1971). 19th Annual Conference on Mass Spectrometry and Allied Topics, Atlanta. Paper F2, p. 85.
8. Eichelberger J. W., Harris L. E. and Budde W. L. (1974). *Analyt. Chem.* **46**, 227.

9. (a) Hertz H. S., Hites R. A. and Biemann K. (1971). *Analyt. Chem.* **43**, 681.
   (b) Costello C. E., Hertz H. S., Sakai T. and Biemann K. (1974). *Clin. Chem.* **20**, 255.
10. (a) Haber J. (1972). M.Sc. thesis, University of California, Los Angeles.
    (b) Haber J. and Jenden D. J. (1973). 21st Annual Conference on Mass Spectrometry and Allied Topics, San Francisco. Paper D5, p. 78.
11. Davis J. E., Shepard A., Stanford N. and Rogers L. B. (1974). *Analyt. Chem.* **46**, 821.
12. Ritter G. L., Lowry S. R., Isenhour T. L. and Wilkins C. L. (1976). *Analyt. Chem.* **48**, 591.
13. Halket J. McK. and Reed R. I. (1975). *Org. Mass Spectrom.* **10**, 808.
14. Simonds J. L. (1963). *J. opt. Soc. Am.* **53**, 968.
15. Rozett R. W. and McLaughlin Petersen E. (1975). *Analyt. Chem.* **47**, 1301.
16. Rosenthal D. and Bursey J. T. (1972). 20th Annual Conference on Mass Spectrometry and Allied Topics, Dallas. Paper T4, p. 419.
17. Carrick A. and Glockling F. (1967). *J. chem. Soc.* (A), 40.
18. Lee J. D. (1973). *Talanta* **20**, 1029.
19. Varmuza K. and Krenmayr P. (1972). *Monats. für Chemie* **103**, 1055.
20. Budde W. L. (1974). Environmental Protection Agency memo. National Environmental Research Center, Cincinnati, Ohio.
21. Bell H. M. (1974). *J. chem. Educ.* **51**, 548.
22. McLaughlin E. and Rozett R. W. (1973). *J. organometal Chem.* **52**, 261.
23. Crawford L. R. (1972/3). *Int. J. Mass Spectrom. Ion Phys.* **10**, 279.
24. McLaughlin E., Hall L. H. and Rozett R. W. (1973). *J. Phys. Chem.* **77**, 2984.
25. Rozett R. W. (1974). *Analyt. Chem.* **46**, 2085.
26. Boone B., Mitchum R. K. and Scheppele S. E. (1970). *Int. J. Mass Spectrom. Ion Phys.* **5**, 21.
27. Braumann J. I. (1966). *Analyt. Chem.* **38**, 607.
28. Evans J. E. and Jurinski N. B. (1975). *Analyt. Chem.* **47**, 961.
29. Semenov V. A., Simonov V. D. and Gudoshnikov S. K. (1974). *Analyt. Chem. U.S.S.R.* **29**, 1414.
30. Shrader S. R. (1971). "Introductory Mass Spectrometry", p. 233. Allyn and Bacon, New York.
31. Burlingame A. L. (1968). In "Advances in Mass Spectrometry" (Ed. Kendrick E.), Vol. 4, p. 15. Institute of Petroleum, London.
32. Beynon J. H. and Williams A. E. (1963). "Mass and Abundance Tables for use in Mass Spectrometry". Elsevier, Amsterdam.
33. Tunnicliff D. D., Wadsworth P. A. and Schissler D. O. (1965). *Analyt. Chem.* **37**, 543.
34. (a) Lederberg J. (1964). "Computation of Molecular Formulas for Mass Spectrometry". Holden-Day, San Francisco.
    (b) Budzikiewicz H., Djerassi C. and Williams D. H. (1964). "Structure Elucidation of Natural Products by Mass Spectrometry", Vol. 2. Holden-Day, San Francisco.
35. Mass Defect Tables (1973). A.E.I. MS 50 Mass Spectrometer Instruction Manual —Operating Instructions. Kratos-AEI Scientific Instruments Ltd., Manchester.
36. Biemann K. and Fennessey P. V. (1967). *Chimia* **21**, 226.
37. Hilmer R. M. and Taylor J. W. (1974). *Analyt. Chem.* **46**, 1038.
38. Hilmer R. M. and Taylor J. W. (1973). *Analyt. Chem.* **45**, 1031.
39. Kunderd A., Spencer R. B. and Budde W. L. (1971). *Analyt. Chem.* **43**, 1086.

5
Library Search

5.1 INTRODUCTION

This chapter is concerned with a discussion of methods in which a file or library of encoded reference spectra is searched to find the best match with a similarly encoded unknown spectrum. The simplest aim of a library search system is to retrieve an identical spectrum from the reference spectra. However, any practical system should be able to retrieve spectra of the same compound although distorted by instrumental variations and many library search systems will retrieve the spectrum of a structurally similar compound if the spectrum of the unknown is not in the library.

This ability to pick out structurally similar compounds can be emphasized by encoding structurally significant features as determined by a knowledge of fragmentation processes rather than simply encoding the most intense peaks. Obviously, a choice of encoding methods is available and these will be discussed in more detail under the section on abbreviation and encoding (Section 5.3). Although library search methods are here discussed separately from interpretative methods (Chapter 7), in which the body of reference data has been formed into rules to allow the interpretation of unknown spectra, any attempted distinction between the methods, which states that library searching cannot be an interpretative routine, is not valid.

At the opposite extreme from the interpretative approach, attempts have been made to maximize the certainty of identification of an unknown which is definitely contained in the library. In this case, the so-called "reverse search" has been applied with some success (1). In the reverse search, the question is asked "is this unknown a particular compound X?" and therefore the library spectrum of compound X is used as the basis of comparison with the unknown spectra. The major advantage of this approach over the conventional or "forward" search is that the encoding of the spectra may be optimized so as to provide the best conditions for a confident identification of a compound, even when the spectrum is quite severely distorted, e.g. by background or unresolved components (1). In its simplest form, the reverse search library contains one spectrum, but obviously each unknown may be compared with the spectra of a number of suspected compounds, perhaps chosen on the basis of gas chromatographic (GC) retention time. In any case, the relatively small library means that the search is very rapid and is suited to real time operation (19d).

Another approach to a more rapid search may be made via the use of an inverted file. With this type of organization, the library file is ordered using one or more important properties such as molecular weight (mol. wt), so that the search can then use these index properties to select certain spectra from the file. This method is to be compared with the commonly used serial or sequential search in which the unknown is compared with each member of the reference file. A good example of the use of an inverted file is the mass spectral search system (MSSS) described by Heller *et al.* (2). This is an interactive system operated on a time-shared computer network with a data file indexed on features such as the presence of a peak at individual mass numbers, mol. wt and molecular formula. The principal method of operation is for the user to specify mass numbers at which peaks should be present, within intensity limits. Thus, to find spectra containing peaks at two specified mass numbers requires only the merging of two lists to locate common entries.

The maintenance of a library on a large time-shared system can have considerable advantages. Fees paid by users are available to provide resources for correcting and updating the library and there is no real restriction on the storage of very large libraries. In addition, users can be encouraged to add their own spectra to the file by use of the network. However, whilst a fee-paying system is suitable for a straightforward search using a large data base, it is probably not an economical means of supporting either more complex interpretative schemes or smaller, more specialized libraries.

Methods of reducing search times in basically sequential systems are also available. For example, the spectra in the main file can be distributed into subfiles by use of a classification algorithm; an unknown is then identified by applying the classification algorithm to select the appropriate subfile and sequentially searching in this subfile. Any compounds that fail the classification algorithm can be subjected to a subsequent, more detailed search against a full library. Again, a simple filter or pre-search that can be carried out much more quickly than the full match can be used to reduce search times by eliminating grossly dissimilar spectra. Finally, it is useful to define two parameters used to assess the performance of file search systems. These are recall, or the proportion of possible matches which are actually retrieved and reliability (or precision), the proportion of positive predictions that are correct (4a).

The detailed discussion of library search systems that follows has been concentrated on three main aspects, viz. (a) the library data base, (b) the abbreviation and encoding of this data base and of the unknown spectra, and (c) the search and comparison procedures designed to identify the unknown spectrum.

5.2 DATA BASES

Any library search system is only as good as its data base. This statement applies both to the coverage of the data (e.g. are there any spectra relevant to the field of interest?) and to its quality. Obviously, most search systems, other than those that cover more restricted compound types will rely on a few major collections of spectra for their data base.

Fortunately, for some years, organizations such as the American Petroleum Institute (API), the Dow Chemicals Company and the American Society for Testing and Materials (ASTM) have been collecting and publishing standard reference mass spectra. Since 1968, the Mass Spectrometry Data Centre (MSDC), has acted as an international centre for the collection and distribution of low resolution mass spectra using a nucleus of the three collections mentioned, together with other smaller ones (see Fig. 5.1). This MSDC data was also used as a data base for search systems developed at the Environmental Protection Agency (EPA), U.S.A. and National Institutes of Health (NIH), U.S.A. and now combined, with the collaboration of MSDC, into a still developing international Mass Spectral Search System (MSSS). The present MSSS data base contains some 30000–40000 spectra, including spectra acquired by EPA and NIH and spectra from the Wiley Registry of Mass Spectral Data (7). The Wiley Registry contains the nominal mass spectra of 18 806 different compounds together with 5073 duplicate spectra, all of which have been extensively checked for errors.

Details of available data bases can be found in ref. 4b. This reference also lists a number of publications of abbreviated mass spectra, of which the "Eight Peak Index of Mass Spectra" (4c), is the most extensive. The "Eight Peak Index" (2nd edition) contains 31 101 spectra, abbreviated so as to give the eight most intense peaks and their relative intensities together with the intensity of the molecular ion if this is not already included. The "Eight Peak Index" is separately indexed by mol. wt and by base peak. Besides proving to be of inestimable value to mass spectroscopists for manual interpretation of mass spectra, the "Eight Peak Index" has also provided a convenient data base for some library search systems (5a, cf. 5b).

The quality of the data in the data base is of the greatest importance. The data may contain errors such as those due to inaccuracies in measurement of mass or intensity and those introduced in data transcription, including abbreviation and encoding of the original spectra for use in the search program. Impurities in the sample or the background of the mass spectro-

MASS SPECTROMETRY DATA CENTRE   AWRE Aldermaston, Berkshire   MSDC
                                                                 4632
2,2-CI(P-CHLOROPHENYL)-TRICHLOROETHANE   (P,P'-D.D.T.)

FORMULA   C14.H9.CL5
M. WT.   352

INSTRUMENT   ATLAS CH4B
INLET SYSTEM   DIRECT
INLET TEMP.   30 C
SOURCE TEMP.   250 C
ELECTRON VOLTAGE 70 EV

M/E	R I	M/E	R I	M/E	R I
35.0	2.1	92.0	0.8	140.0	0.8	206.0	1.9
36.0	4.9	93.0	1.3	141.0	1.0	207.0	0.7
37.0	0.8	94.0	0.8	146.0	1.2	208.0	1.7
38.0	1.5	97.0	1.5	147.0	1.0	209.0	0.8
39.0	4.0	98.0	3.0	148.0	1.2	210.0	2.8
45.0	0.5	99.0	6.9	149.0	2.8	211.0	1.8
47.0	1.0	100.0	3.7	150.0	3.0	212.0	10.9
49.0	1.0	101.0	5.5	151.0	2.1	213.0	2.1
50.0	12.5	102.0	1.0	152.0	1.2	214.0	3.9
51.0	9.1	105.0	4.9	156.0	0.0	215.0	0.8
52.0	0.8	106.0	1.3	159.0	0.7	235.0	100.0
58.0	1.3	107.0	3.2	160.0	1.0	236.0	15.7
60.0	1.3	108.0	0.8	161.0	0.8	237.0	67.7
61.0	1.9	109.0	1.7	162.0	1.3	238.0	9.5
62.0	3.9	110.0	2.5	163.0	8.0	239.0	11.6
63.0	7.1	111.0	4.0	164.0	8.2	240.0	2.1
64.0	2.6	112.0	1.1	165.0	53.1	246.0	7.4
65.0	0.8	113.0	3.1	166.0	8.0	247.0	2.8
67.0	0.5	114.0	1.4	167.0	1.0	248.0	5.1
68.0	0.5	115.0	1.4	170.0	3.0	249.0	1.3
69.0	1.0	117.0	5.4	170.2	0.0	250.0	1.0
70.0	0.8	119.0	5.4	171.0	3.1	281.0	1.3
72.0	2.6	121.0	1.4	172.0	2.3	282.0	3.0
73.0	4.0	122.0	1.4	173.0	3.4	283.0	1.9
74.0	9.0	123.0	8.6	174.0	2.8	284.0	3.0
75.0	20.1	124.0	3.4	175.0	2.8	285.0	1.0
76.0	4.0	125.0	2.3	176.0	11.8	286.0	1.0
81.0	1.3	126.0	2.2	177.0	3.7	317.0	1.7
82.0	7.4	127.0	0.8	178.0	0.8	319.0	1.0
84.0	2.1	134.0	3.7	184.0	1.0	321.0	1.0
85.0	3.4	135.0	9.6	185.0	0.8	352.0	2.4
86.0	3.4	136.0	0.0	186.0	1.0	354.0	3.4
87.0	5.9	137.0	3.8	199.0	16.5	356.0	2.5
88.0	7.0	138.0	4.9	200.0	9.3	358.0	0.8
89.0	2.6	139.0	2.8	201.0	6.6		
				202.0	3.7		

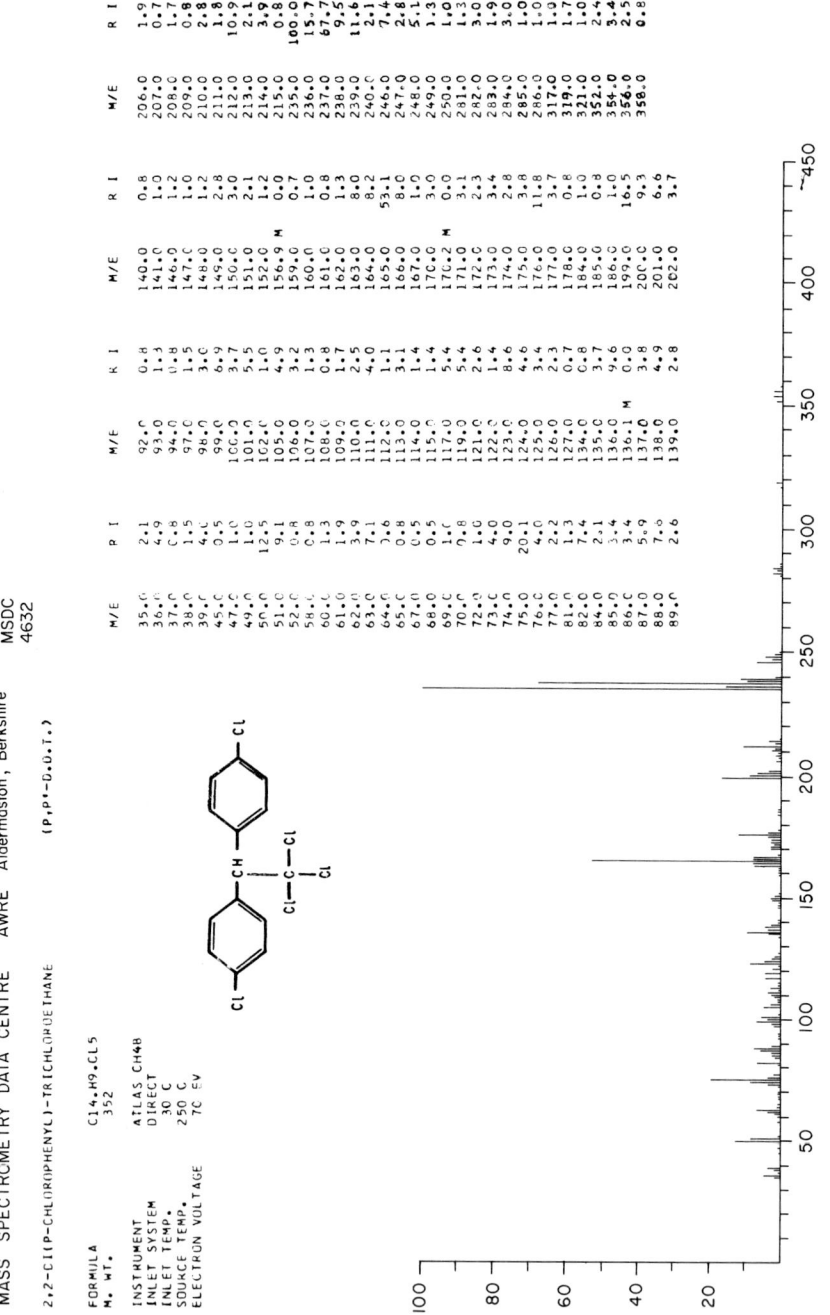

Fig. 5.1. Information from a typical Mass Spectrometry Data Centre (MSDC) data sheet.

meter will produce extraneous peaks in the mass spectra. Differences in mass discrimination between different instruments, variations in the mass and intensity ranges recorded temperature effects in the inlet system or ion source, ion-molecule reactions and memory effects can all cause considerable variations in recorded spectra. However, despite this apparently overwhelming list of sources of non-reproducibility, library search techniques can work very successfully. Part of the reason for this success lies in the use of encoding and comparison techniques that can minimize the effect of this variability (6).

In gas chromatography–mass spectrometry (GCMS) work, the rapidly changing sample concentration, even during the course of a fast scan, the background due to column bleed which is particularly evident in the analysis of small amounts of material and in analyses at elevated temperatures, and the contribution to spectra from unresolved components of the mixture, all serve to lower still further the reproducibility of spectra recorded under these conditions. In the earlier days of mass spectrometry, standard spectra used in the quantitative analysis of mixtures, particularly in the petroleum industry, were recorded under very carefully controlled and calibrated source conditions and in many cases the same type or even the same instrument was always used for this work. Whilst this approach is generally not possible in GCMS analysis, it is worth pointing out that spectra recorded on a user's own instrument and under standard conditions are to be preferred to any others in a library. Thus, updating a library with a user's own spectra is to be recommended, although these spectra should undergo the same thorough checks for errors and inconsistencies as would spectra from an external source.

The frequency of occurrence of errors in published data is well illustrated by figures given in the Preface to the Wiley Registry of Mass Spectral Data (7). In checking the 18806 spectra for the Registry, probably 10000 changes and corrections to the published data were made and as the editors point out "the very magnitude of this figure suggests that some gross errors must still be present". The Preface also describes a program that can check for three common symptons of poor data: (a) impurity peaks at masses above the molecular ion, (b) isotopic abundance ratios that do not correspond to those expected from the elemental composition, and (c) the presence of illogical neutral losses (8). As is also pointed out, the problem of correcting such errors or choosing among duplicate spectra, where they exist, is often more difficult than that of locating these errors. Some criteria for choosing among duplicate spectra were suggested. In addition to criteria based on the presence or absence of the error conditions listed above, decisions were made on the basis of the total number of peaks, the mass discrimination (i.e. spectra with more intense higher mass peaks were preferred) and the source from which the spectra were obtained.

5.3 ABBREVIATION AND ENCODING

Selection and encoding of certain mass peaks or features from a spectrum reduces both the storage and time requirements of a search system. Retaining a complete spectrum including intensity values presents an inordinate amount of data for storage and searching except for very small, specialized libraries (9). For example, while it may be possible to store perhaps 4000 full spectra on a 2M word disc, this number can be increased by a factor of 10 or more depending on the encoding method chosen. In addition, substantial abbreviation will allow a small working library to be stored in core for real time operation.

However, abbreviation fulfils another equally important task, that of improving the performance of the search by minimizing the variability of recorded spectra. For example, use of a minimum intensity threshold for encoding ensures that low intensity peaks of doubtful reproducibility or relevance are not used in the matching procedure. Fortunately, in this respect, the great majority of nominal mass spectra are over-determined, that is to say they contain more information than is necessary to provide a unique identification of the compound. Thus, information may be discarded without impairing the performance of the retrieval system and by careful selection of data for encoding, the performance will actually be improved. For example, Mathews and Morrison (10) showed that the matching of six to eight major peaks in terpene spectra gave results comparable to the use of full spectra. Again, the encoding of intensity values should not be so tightly defined that spectra of the unknown compound recorded under different conditions are rejected in the search.

Any practical encoding system, particularly for application in GCMS work, must be able to make due allowance for background peaks. Known major background peaks, e.g. m/e 207 from silicone column bleed, may be excluded from consideration altogether and normalization only carried out with respect to a base peak with an m/e value greater than 44 so that very intense background peaks from air such as m/e 28 (N_2) do not distort the spectrum. In any conventional forward library search routine applied to GCMS work, a complete background subtraction or background correction (cf. Section 4.4) should always form the first stage of the search.

As the strength of any library search system lies in the quality of its data base, it is also important that the selection and encoding of newly recorded spectra that may serve as reference spectra should be as simple as possible. Regular updating of the data base will be encouraged in this way.

The choice of an encoding system can be made in terms of three basic systems, viz. (a) recording of the intensity of a limited number of peaks, (b) transformation of the spectrum into "features", e.g. intensity ratios at selected masses, the presence of peaks at certain masses or of certain neutral losses in the spectrum and (c) a mathematical function calculated using the whole spectrum.

5.3.1 Encoding of a Limited Number of Peaks

The simplest and one of the earliest methods used was to encode the masses and intensities of only the N most intense peaks (11, 1, 5b, 6a) (See Fig. 5.2). The value of N used varied between 5 and 10 but could conveniently take a value that corresponds to published compilations of abbreviated spectral data, e.g. the "Eight Peak Index" or the 10 peak index of Cornu and Massot (12). The method is simple and direct both in the encoding and the subsequent matching. The method also allows a surprisingly unique description. For example, a library of 65 compounds, including 58 commonly used drugs, was uniquely described by a selection of the m/e values of only the five most intense peaks in the mass spectrum, listed in order of decreasing abundance (11a).

However, the method does have some drawbacks. Investigations have shown that the method is less satisfactory for general libraries containing many compounds with mol. wt greater than 200 (6a, 11c). This is because a choice of the N most intense peaks in many cases results in selection of only the intense, but structurally less relevant, low mass peaks from a spectrum. Modifications which could improve the performance with higher mol. wt compounds to some extent include the use of a lower mass limit, e.g. the 10 most intense peaks above m/e 40, and weighting the higher mass peaks (6a), for example, by use of a weighted intensity, $I' = I \times m/e$ (11e). The other serious drawback of such a system is that, as Knock et al. (6a) point out, it takes no account of possible mass discrimination effects due to the type of instrument used. A later investigation by McGuire et al. (6b) on the design of a search system that would compare unknown spectra from quadrupole instruments, where the spectra exhibit a bias towards a greater intensity at low mass, with a library mainly comprising data from magnetic sector instruments, concluded that use of the N most intense peaks was not suitable. The investigations of Knock et al. (6a) also suggested that a system using the N most intense peaks was less satisfactory with thermolabile compounds exhibiting extreme variability of spectra although, in the example chosen, the sesquiterpene farnesol, much improved results were obtained when N was reduced to a value of 5 or 6 from the original 15.

In summary, the method is a very simple one which may be entirely suitable for compounds of lower mol. wt particularly where libraries of a more limited size are employed. In addition, the application of the method in conjunction with a small MS of limited resolution and scanning only a restricted mass range, should be noted (11d). In this work, simple organic compounds were identified by use of the six most intense peaks in the range m/e 150–25 taken from scans that covered only this mass range.

The natural development of the use of the N most intense peaks has been the encoding of n peaks every m mass units, e.g. 2 peaks every 14 mass units (see Fig. 5.2). This method does not abbreviate the spectrum to the same extent as the N most intense method, but it does overcome the two principal objections to that method, viz. poor results with higher mol. wt compounds and inapplicability to quadrupole or other spectra showing mass discrimination. The earliest published records of this method are from Petterson and Ryhage (11c) who suggested the use of two masses every 25 a.m.u. and from Abrahamsson et al. (11e). However, the best known and most used method is that of Hites and Biemann (14) which employs the two most intense masses every 14 a.m.u. The choice of a 14 mass unit interval is more logical interpretatively because it ensures that significant peaks belonging to a homologous series of ions are selected if they are relatively abundant. In addition, this technique necessarily includes the molecular ion region since it, and the heaviest fragment ion cannot possibly be in the same group, except for ions due to the loss of hydrogen. Originally, the intervals m/e 1–14, 15–28, etc., were used for encoding (14). However, these were changed to m/e 6–19, 20–33, etc., when it was realized that difficulties occur if the boundaries split common peak clusters, e.g. m/e 42 and 43 (13a).

A number of other authors have used division into 14 a.m.u. intervals, but have only recorded the most intense peak in this interval (15). An example of the use of one peak every 14 mass units is the encoding of the spectra of 133 drugs of abuse by Finkle et al. (15c). For this library, the previously described system due to Fales (p. 107) (11a) using only the five most intense peaks proved unable to discriminate between some of the isomeric drugs. Conversely, the encoding and comparison techniques of Hites and Biemann (13a, 14) proved unnecessarily sophisticated for this limited application. However, as with the N largest peaks system, the system of Finkle et al. benefited from the use of a lower mass limit for encoding (m/e 34) to eliminate background and from confining the selection of the base peak to peaks of mass greater than m/e 48 to eliminate interferences such as carbon dioxide at m/e 44 and argon at m/e 40. Equally important, it was necessary to have a rule to decide which peak to encode if it was not possible to determine which was the most intense peak in any group of 14 a.m.u. In this case the peak of higher mass was arbitrarily chosen. Grotch (15a) has shown that a similar approach, where the higher mass peak rather than the most intense was encoded if the intensity ratio of the two most intense peaks was less than 1·20, improves the reliability of identification. In view of the difficulty of obtaining reproducible intensity data in practice, this result is perhaps not surprising.

Robertson and Merritt (16) proposed a simple alternative scheme in which the most intense peak in every 7 a.m.u., interval beginning at m/e 23, is encoded. The position of the peak within each interval is specified by three bits, hence the term "octal coding" used to describe this system. Five such peaks can be encoded in a 16 bit word; the remaining bit serves as a final word flag (see p. 139). The authors suggest that octal coding provides a successful basis for the identification of functional group character when the unknown compound is not in the library (see Table 5.1). A system described by McLafferty as part of the Self-training Interpretative and Retrieval System (STIRS) provides a further variation by encoding the most intense odd mass peak and the most intense even mass peak in each 14 a.m.u. interval (23). The system described by Bell (17) encodes the three most intense peaks in each 14 a.m.u. interval.

TABLE 5.1
Examples of octal coding[a]

Pattern for alcohols				Pattern for esters			
Compd.				Compd.			
A	7272			A	7202	03	
B	5272	02		B	7271	0203	
C	7262	72		C	7272	04620	3
D	7272	52		D	7272	04620	3

Pattern for aldehydes				Pattern for aromatics			
Compd.				Compd.			
A	7131	7		A	5037	72607	10000
B	7271	7071		B	5037	71616	06100
C	7251	71717	1	C	5037	16161	60710
D	7051	71707	1	D	5037	16161	60607	1

[a] Alcohols, A–D are ethanol, iso-propanol, n-propanol, 2-ethyl-l-butanol; aldehydes; C–D are two different spectra for 3-methylbutanal; aromatics, A–D are benzene, toluene, ethylbenzene and propylbenzene. (From ref. 16a.)

The successful use of a system of coding of the most intense peak or peaks from successive mass intervals by a large number of authors (18, 11c, 11e, 13, 14, 15, 16, 17) in practical situations is ample testimony to its value. Apart from its general success, McGuire et al. (6b) have shown that this form of encoding is able to make successful library comparisons of quad-

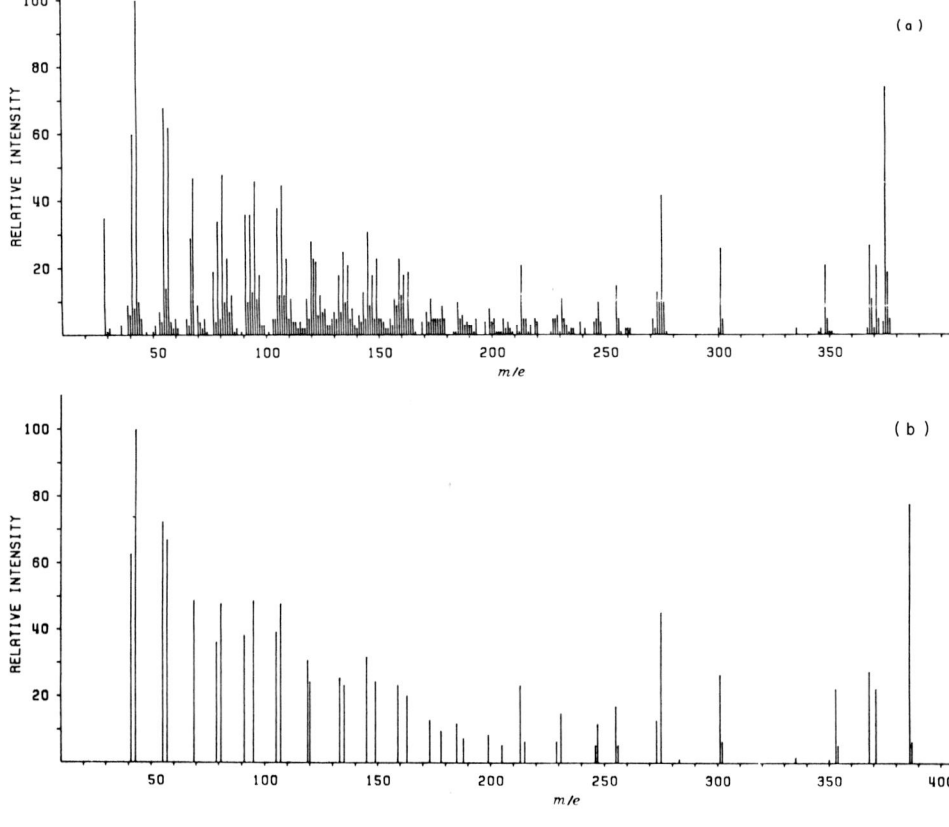

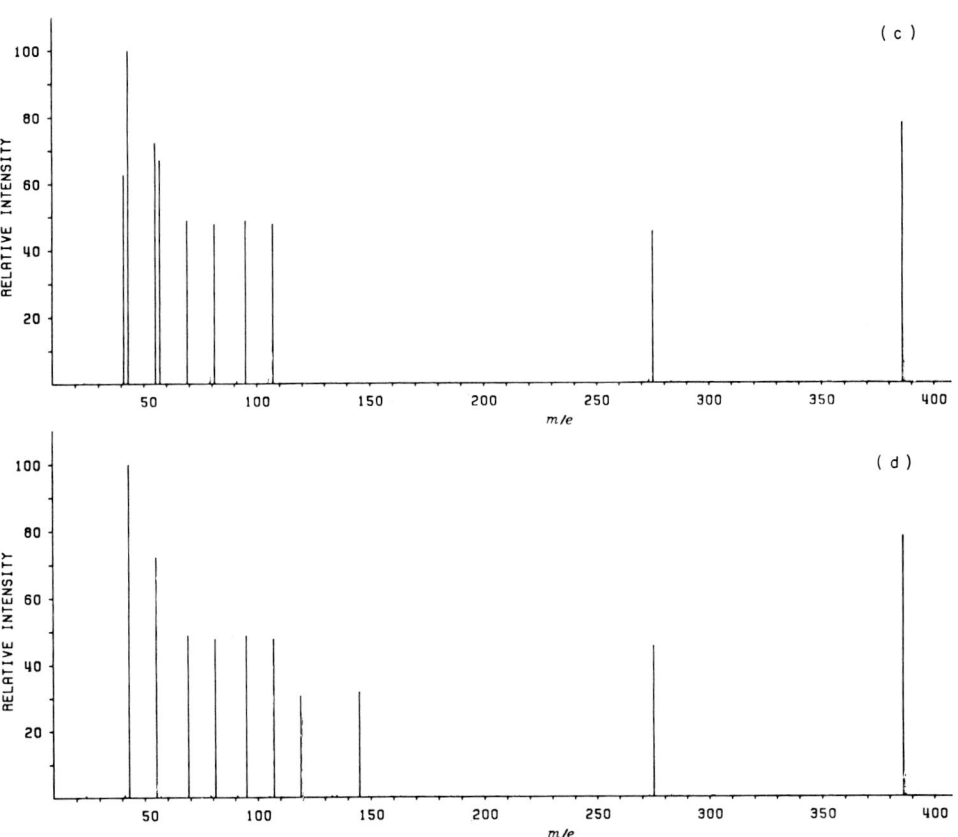

Fig. 5.2. Mass spectrum of cholesterol illustrating various abbreviation methods. (a) Complete spectrum. (b) Two most intense peaks every 14 a.m.u. (c) 10 most intense peaks. (d) 10 most "significant" peaks using algorithm of Grönneberg et al. (19b). (Adapted from ref. 3.)

rupole and magnetic sector instrument data (cf. p. 107). It has also been shown by Knock *et al.* (6a) to be a markedly more successful method than the use of the N most intense peaks in matching the spectra of thermolabile compounds. One disadvantage of encoding the most intense peak or peaks in successive mass intervals is that the method produces variable record lengths depending on the mol. wt of the compound.

Another means of obtaining a restricted number of masses to represent a spectrum has been to choose the N most significant peaks (1, 19). This choice has often been carried out manually on entering a compound into the library and the peaks chosen have been those assumed to be of greatest structural relevance. This system has usually been employed with libraries restricted in size and/or type of compound so that the peaks chosen are for a somewhat more specific comparison than a general search of, say, a 10000 spectrum library. In particular, the method has been employed with the reverse search in which evidence for the presence of a particular compound is sought rather than comparing the unknown with a full library to find the closest match. As Sweeley and coworkers (19c) have pointed out, in this situation optimum specificity is obtained by selecting ions which are not only characteristic of the compound sought, but which also differentiate it from other compounds likely to be present with that compound, e.g. compounds which are not well separated in GCMS.

A method of choosing the N most significant peaks that can be implemented automatically has been suggested by Grönneberg *et al.* (19b). Starting at highest mass, only those peaks whose intensities exceed half the average intensity of peaks in the spectrum are collected. If a group of such peaks occurs, differing by only 1 or 2 mass units, then only the most intense peak in the group is selected. The collection of peaks is continued until either all peaks have been considered or at least N peaks have been selected and at least half the sum of the total ion current in the spectrum has been accounted for. Finally, the N most intense of the selected peaks are used (see Fig. 5.2). Another method used in the probability based matching (PBM) system developed by McLafferty (19e) basically chooses those peaks with the highest $(U + A)$ values as being the most significant (see p. 136). The use of about 10 significant peaks is usual in most published methods. The number of published applications of the use of the N most significant peaks is presently small but is likely to increase with the more widespread use of restricted libraries and the reverse search principle.

A different approach to peak selection for encoding has been described by Wangen *et al.* (20). They began their investigations by encoding the presence or absence of a peak above a 1% threshold at each mass position using a binary code of 352 bits. However, they observed that with a binary code, the information per channel (mass position) is given by:

$$h = -[p \log_2 p + (1 - p) \log_2(1 - p)]; 0 \leqslant p \leqslant 1 \qquad (5.1)$$

where p is the probability of the given mass position containing a peak, i.e. the fraction of spectra having the given mass fragment. The total information is obtained by summing over all masses and is at a maximum when every value of p is 0·5, implying that the information content will be maximized when all bits are on for half the compounds in the library. Most bits are on for far less than 50% of the compounds, thus it was decided that a combination of various mass channels, with the sole condition that the sum of the individual occupancy levels is approx. 50%, might be able to effect a significant reduction in dimensionality without a drastic loss of information content. A first reduction from 352 to 80 bits was effected by an arbitrary combination of mass positions to 50% occupancy and then a reduction to 48 bits by preferentially combining mass positions that correlate well, i.e. occur together in spectra. An analysis of the library showed that the best correlations were obtained for positions separated by 1, 2, 13, 14 and 15 mass units. The results using 48 bits were generally good and were comparable with those obtained using 352 bits, thereby considerably reducing storage requirements and greatly increasing search speeds. In addition, the best matches, apart from the unknown compound itself, were often structurally more closely related with the 48 bit coding than with the 352 bit coding. This seemed to be an effect of combining highly correlating mass positions. It would seem that further investigations on the lines suggested by these authors should prove extremely fruitful.

A similar approach to peak selection has been described more recently by van Marlen and Dijkstra (21). Again working with binary encoded spectra and using an encoding threshold of 1%, these authors calculated the information content of each mass position for a set of approx. 10 000 spectra. Further analysis allowed the determination of an optimum sequence of 120 masses which contained practically all the information (approx. 40 bits) from the original 300 or so masses. The optimum sequence begins with the mass that yields the highest amount of information and subsequent masses are chosen, using a covariance analysis, so that the information content of the combined masses is maximized. The use of a covariance analysis parallels the consideration of correlating mass positions by Wangen et al. (20).

Van Marlen and Dijkstra also make a number of important points in their paper (21). Firstly, the peaks selected as containing maximum information with respect to retrieval will not necessarily have any major significance as regards interpretation of mass spectra. Secondly, the amounts of information are strongly influenced by the choice of threshold level. Undoubtedly, optimum thresholds could be derived for each mass rather than the general 1% threshold used in this and other studies (20, 22).

5.3.2 Encoding as Features

Encoding of spectra as features has been principally carried out in two large library search systems, the Self-training Interpretative and Retrieval System (STIRS) described by McLafferty (23) and the system described by Clerc et al (24). As an example, the Clerc system, which employs binary coding to indicate the presence or absence of each feature in a spectrum, codes the following general types of features:

(a) The presence of a peak at $m/e = x$ if the intensity at that mass exceeds a selected threshold value.
(b) The occurrence of losses of n mass units between significant peaks in the mass spectrum.
(c) The presence of ion series summed every 14 mass units (e.g. m/e 15, 29, 43, 57, 71, etc.) with a total intensity exceeding a selected value.
(d) The presence of ion series summed every two mass units (i.e. odd mass series and even mass series) with a total intensity exceeding a selected threshold value.
(e) An approximate estimate of molecular size.
(f) A comparison of the fraction of the TIC in each third of the spectrum mass range with a selected threshold value.

The STIRS system encodes similar features, but on a different basis. For example, it is, within limits, the most intense masses, or ion series, that are encoded rather than the presence or absence of specific masses or ion series. In both these systems, a knowledge of mass spectral fragmentation behaviour is used initially to identify for the computer the features that should be of greatest structural significance. To the extent that these spectral features reflect particular structural features, e.g. functional groups, skeletal type, molecular size; these systems should be able to characterize the unknown by selecting best matches containing structural features that are the same as or similar to those in the unknown. This facility is especially valuable in a general library search when, as may often happen. the unknown is not in the library. Indeed, the full STIRS system is only employed when a simple retrieval system such as PBM (see p. 135) has failed to produce a match of high reliability (25). In addition a system based on encoding as features should be less affected by instrumental variations or by contaminated spectra.

Specifically, the features that describe the data classes used in the original STIRS system are (a) ion series; (b) low, medium and high mass characteristic ions (3 classes); (c) small and large neutral losses (2 classes); (d) secondary neutral losses (3 classes); (e) fingerprint ions. A final class is an overall match factor derived as a linear combination of classes (a)–(c). On the basis of a statistical assessment of the performance of STIRS with various data classes (see p. 132) later papers from the Cornell group have described improved performance using overlapping mass ranges in the definition of type (b) and (c) classes. They have also shown that the use of overall match factors both for type (b) and (c) classes leads to more reliable substructure prediction than does the use of the individual classes (4a, 26). In addition, some doubt was thrown on the utility of the secondary loss classes used in the original system (27).

As an example, from the mass spectrum of the bis-trimethylsilyl ether of 1-(acetylamino)ethylphosphonic acid, the individual data classes of STIRS predicted the substructures CH_3CONH-, $-Si(CH_3)_3$, P, $-CH_3$, $-NH_2$ and a double bond equivalent of two. For the same unknown, the 15 compounds selected with the highest overall match factor values showed the common substructural features listed in Fig. 5.3.

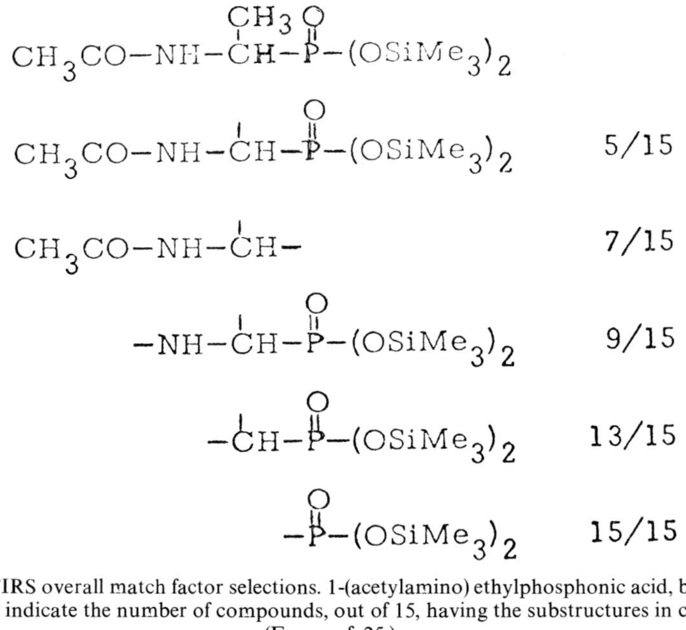

$$CH_3CO-NH-\underset{|}{CH}(CH_3)-\underset{\parallel}{P}(O)-(OSiMe_3)_2$$

$$CH_3CO-NH-\underset{|}{CH}-\underset{\parallel}{P}(O)-(OSiMe_3)_2 \qquad 5/15$$

$$CH_3CO-NH-\underset{|}{CH}- \qquad 7/15$$

$$-NH-\underset{|}{CH}-\underset{\parallel}{P}(O)-(OSiMe_3)_2 \qquad 9/15$$

$$-\underset{|}{CH}-\underset{\parallel}{P}(O)-(OSiMe_3)_2 \qquad 13/15$$

$$-\underset{\parallel}{P}(O)-(OSiMe_3)_2 \qquad 15/15$$

Fig. 5.3. STIRS overall match factor selections. 1-(acetylamino) ethylphosphonic acid, bis-TMS. The figures indicate the number of compounds, out of 15, having the substructures in common. (From ref. 25.)

In his initial review of the performance of the STIRS system, McLafferty (23) remarks on two, somewhat unexpected, features. One is the structural specificity shown by some low mass ions (i.e. $m/e < 90$) and the other is the identification of quite complex structural features from a consideration of major losses from the parent ion. The value of a consideration of neutral losses in interpretation of mass spectra is certainly well known to practising mass spectroscopists. However, their comprehensive derivation from a mass spectrum is not easy. A feature of the STIRS system is the use of a monoisotopic spectrum for derivation of neutral loss data.

TABLE 5.2

Statistics of losses with more than 6500 occurrences in the file of 12 769 spectra[a]

Loss	# Refs
0	10 822
43	7776
29	7581
57	7424
15	7398
71	7072
45	6686
28	6621
1	6595
44	6545
56	6508

[a] From ref. 28.

The MSSS (18c) also allows a search for spectra with losses of 0–150 a.m.u. from the molecular ion and the MSSS manual (28) gives valuable statistics on the occurrence of the most common losses from the molecular ion in a file of 12 769 spectra (Table 5.2). A program developed by Spiteller (29) uses the recognition of key ions and key losses to identify known steroids and provide partial structures of unknown steroids.

5.3.3 Encoding as a Mathematical Function

Robertson and coworkers (16b, 30) used the Khinchine function as a single valued representation of a spectrum which could easily be stored in a data file. The function was calculated from eqn 5.2:

$$-\eta = \sum_{i=1}^{n} p_i \log p_i \qquad (5.2)$$

where p_i's are the normalized intensities at each mass. The function was found to be uniquely representative in better than 90% of the cases investigated. In the remaining cases a divergence function, J (1, 2), was used to differentiate pairs of similar spectra. This function was calculated from eqn 5.3:

$$J(1,2) = N_1 \sum_{i=1}^{n} (p_{1i} - p_i) \ln \frac{p_{1i}}{p_i} + N_2 \sum_{i=1}^{n} (p_{2i} - p_i) \ln \frac{p_{2i}}{p_i} \quad (5.3)$$

where

$$p_i = \frac{p_{1i} + p_{2i}}{2}.$$

One problem with these diagnostic functions was the lengthy calculations needed in their determination. However, more importantly, variations in recorded intensities with the same compound in different MSs produced considerable uncertainty in the reliability of diagnostics using these functions.

Attention may also be drawn to work typified by that of Rabinovich and Sakharov (31) in which the relative intensity of major peaks common to most spectra is used as a diagnostic function. Their work demonstrated a simple

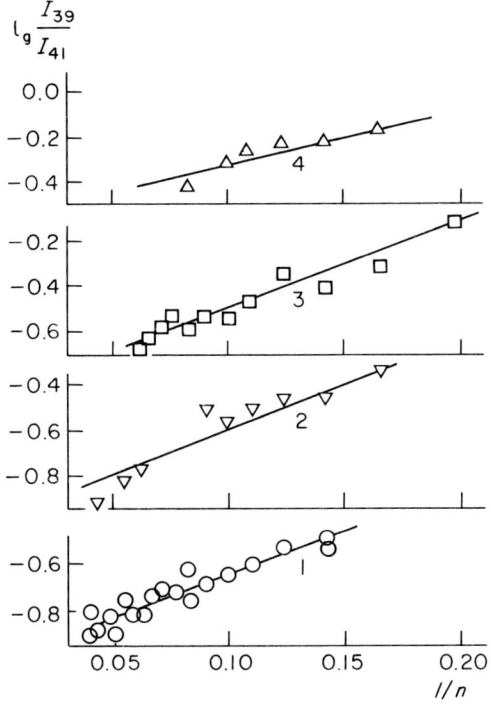

Fig. 5.4. Log I_{39}/I_{41} v. the number n of carbon atoms in homologous series of compounds: (1) n-alkanes; (2) 2-methylalkanes; (3) alk-l-enes; (4) alk-l-ynes. (Reprinted with permission from ref. 31. Copyright by Plenum Press.)

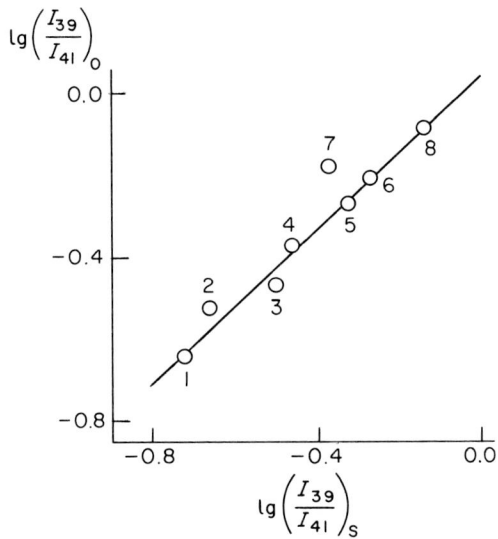

Fig. 5.5. The relationship between the values of $\log(I_{39}/I_{41})$ for oxygen and sulphur containing compounds: (1) methoxyethane—methylthioethane; (2) butanol-1—butanethiol-1; (3) 2-methylpropanol-2—2-methylthiopropane; (4) hexanol-1—hexanethiol-1; (5) ethyl isobutyl ether—5-methyl-3-thiohexane; (6) 1-methoxybutane—2-thiopentane; (7) propanol-1—propanethiol-1: (8) pentanol-2—pentanethiol-2. (Reprinted with permission from ref. 31. Copyright by Plenum.)

linear dependence of the ratio $\log(I_{39}/I_{41})$ (I_m = intensity at mass m) when plotted against the reciprocal of the carbon number in homologous series or against the degree of unsaturation in compounds with the same carbon number (Fig. 5.4). In addition, a simple linear relationship between the corresponding values of $\log(I_{39}/I_{41})$ was established when oxygen was substituted by sulphur in a number of aliphatic compounds (Fig. 5.5). Whilst this particular approach to structural identification did not involve the use of a data system, it does serve to illustrate the wide range of structural data which is present in and may be simply encoded from low resolution mass spectra.

5.3.4 Use of Intensity Data in Encoding

Possibly the simplest means of making use of intensity data is to use it in a relative manner (Section 5.4.2). An example is given in the work of Knock *et al.* (6a) (Method 2) which is used in the "Massmatch" system available from the MSDC. In this the N most intense peaks in both the unknown and library spectra are arranged in order of decreasing intensity and the relative positions of peaks with the same m/e values compared. No absolute values of intensity are used (see eqn 5.8 and Fig. 5.8). This method affords an economical means of making use of intensity data and a rapid search routine.

Binary representation is another economical means of encoding and lends itself to very rapid processing (Section 5.4.1). It also provides a surprisingly unambiguous representation. For example, in encoding 3246 spectra to one bit/mass over the mass range 13–140, Grotch found only 15 pairs of identical binary spectra, seven of which were due to structural isomers (22). In studies on the binary encoding of full spectra, it was found that encoding above a threshold of 1% base peak intensity caused very little loss of information compared with the use of lower thresholds (22, 20). The use of a relatively high threshold such as 1% is, in any case, necessary to try to avoid the coding of background peaks along with true sample peaks. Unfortunately, coding of background peaks using this method remains a serious problem, particularly when dealing with low level components eluted from the gas chromatograph (13a). Again, Mathews and Morrison (10) found that one bit matching of terpene spectra gave significantly poorer results than other methods using more intensity information. These authors found that one bit encoding of the terpene spectra was most successful when a threshold level of about 10% of base peak intensity was used, suggesting that the lower intensity peaks did not contain specific information which could distinguish these closely related spectra.

The addition of some intensity data in the encoding of full spectra gives better matching results than binary coding in so far as the correct structure achieves a higher ranking in the list of best matches (32). However, this improvement is necessarily at the expense of increased storage and search time. The information encoded by this method is maximized if each intensity level is equally populated (cf. p. 113) and, as peak heights in mass spectra are distributed log-normally (33, 40), this can be effected by choosing a logarithmic scale of transitions between levels. In this case, a three bit encoding (i.e. eight levels) of intensity with transitions at 0·5, 1, 2, 4, 8, 16 and 32% base peak intensity was used. Further improvements could be obtained using the full intensity range from the minimum recorded intensity of 0·01% base peak stored in 32 bits. Once again, this improvement was only achieved at the expense of slower speeds (search rates were an order of magnitude slower than in the corresponding one bit case) and greatly increased data input.

In studies on the encoding of only the most intense peak in 14 a.m.u. intervals (15a) it was again found that improved results were obtained by the addition of some intensity information. In this case, two bit encoding (four levels) was used, also with logarithmically distributed transition levels (0·01, 4·2 and 22·4% base peak intensity). This particular combination of mass and intensity data actually gave better results than the use of the full spectra. The same form of encoding was used by Markey and coworkers in their studies on metabolic profile analysis using GCMS techniques (18a, 34).

As a general point, it should perhaps be mentioned again that in all these examples of the use of intensity data, where two peaks have intensities that differ by a factor less than, say 1·20, the preferred method is to arbitrarily encode the higher mass peak rather than the more intense peak (cf. p. 108).

5.4 SEARCH AND COMPARISON ROUTINES

The sections describing individual search and comparison routines attempt to give an idea of the different methods used. These can be described under the general headings of (a) logical operator comparisons, (b) comparisons based on relative peak intensities, and (c) comparisons based on absolute peak intensities. In order to reduce costs, most search routines make use of a filter which is either a pre-search or is built into the search itself, as in a search using an inverted file. The filter eliminates grossly dissimilar spectra so that a matching score is only computed for those pairs of spectra that have some prospect of matching.

A typical output from a search consists of the M best matching compounds with their respective scores or measures of similarity, where M is typically 6–10 for a forward search and a much smaller number, perhaps 1–2 for a reverse search. Another possible output, viz. all reference spectra with a matching index $\geqslant R$ (20), could be used and might be faster in operation since new matches need not be compared to the existing M best for updating, but only to the value of R. However, the difficulty in defining a suitable value of R for all searches and the production of variable length lists has meant that this idea has not been generally adopted. In most cases, a final step consists of normalizing the computed match score to provide a proper basis for comparison of matches. A score of one for a perfect match and zero for an absolute mismatch is a common convention.

5.4.1 Logical Operator Comparisons

Computer hardware is most efficiently utilized by performing operations on "word" size bit strings. This is easily achieved using one bit encoding in which case the ones and zeros, which may represent the presence of absence of a peak at each mass above the threshold, packed into computer words, are subject to logical operations on a word by word basis. In this way, it is possible to achieve parallel processing (i.e. simultaneous channel by channel comparisons) of the data with extremely rapid search speeds as a result. As an example of this technique, an earlier method described by Grotch (22) employing one bit encoding of the intensity at each mass, uses the "logical exclusive or" (XOR) instruction to compare the unknown and library spectra. Figure 5.6 illustrates the comparison of two words corresponding to two encoded spectra, A and B, and the resultant word C, where $C = \text{XOR}(A, B)$. As can be seen, XOR is a measure of distance or disagreement and the total number of mismatches may be obtained by summing the ones over all the channels in C.

WORD	1	2	3	4	5	6	7	....	M
A	0	1	1	0	1	1	0	....	1
B	1	1	0	0	1	0	1	....	1
C	1	0	1	0	0	1	1	....	0
D	0	1	0	0	1	0	0	....	1

BIT POSITION ⟶

Fig. 5.6. Examples of "logical exclusive or" (XOR) operation, $C = \text{XOR }(A, B)$ and "logical and" operation, $D = \text{AND }(A, B)$. M = number of bits in a computer word. (Adapted from ref. 22.)

A similar measure of disagreement based on calculation of XOR has been used by Wangen et al. (20). However, as Grotch (32) later pointed out, the use of XOR alone is intuitively unsatisfactory as it gives equal weighting to the case where peaks are present at one mass in both spectra and to the case where peaks are absent in both spectra. To try to overcome this deficiency a composite disagreement function XOR-μ (AND), where μ is a weighting factor for "logical and", was used in place of XOR (32) (see Fig. 5.6). Thus, the case where peaks are present at one mass in both spectra is taken to show better agreement than the case where they are both absent. The actual criterion to be minimized was:

$$\mu N + \sum_{i=1}^{M} [(\text{XOR})_i - \mu(\text{AND})_i] \quad (5.4)$$

where M is the number of masses to be compared, N is the number of peaks (ones) in the unknown and μ was found experimentally to have an optimum value of two. This particular expression uses an offset of μN to ensure that the minimum value of the criterion is zero. This same expression was used by Hardy and Jardine ($\mu = 2 \cdot 3$) (18b) and by Merritt et al. ($\mu = 2$) (16b) in matching binary spectra.

Subsequent work by Grotch (15a, 35), using a binary representation of the most intense peak every 14 a.m.u. showed a steady improvement in the search results through refinement of the matching criterion. The first change (15a) was to use a variable value of μ where:

$$\mu = 3 + \log_{10}(\text{intensity}); \ 1 \leqslant \mu \leqslant 5 \tag{5.5}$$

so that more weight was given to results from more intense peaks. In a further development, Grotch (35) employed a method based on hypothesis testing to calculate weights which optimize matching performance. Weights, characteristic of the library in question, are calculated for each possible combination of binary data for each mass channel n. They are referred to as $\alpha_n(00), \alpha_n(01), \alpha_n(10)$ and $\alpha_n(11)$. The dissimilarity index, d, is calculated as

$$d = \sum_{n=1}^{N} \alpha_n(ij) (i = 0, 1 \ j = 0, 1 \text{ in the binary case})$$

and is then normalized so that the dissimilarity index, d^*, becomes:

$$d^* = \frac{d - d_{\min}}{d_{\max} - d_{\min}} \tag{5.6}$$

where $d_{\min}$ and $d_{\max}$ are the values of d for a perfect match and complete mismatch respectively. Now d^* has a range of values from zero for a perfect match to one for a complete mismatch.

Grotch developed formulae that allow the calculation of these weights by pairwise examination of all redundant spectra in the library being used. These redundant spectra are spectra of the same compound that differ only because of errors of observation or encoding in either or both. Thus, statistics developed from these spectra allow calculation of weights that minimize the separation of such spectra but maximize the separation of truly dissimilar spectra in that library. From these weights, an average optimum value of $\mu \ (\mu = -\alpha(11)/\alpha(10))$ can be calculated for the library. For the case described it is 1·6, acceptably close to the value of $\mu = 2$ previously found effective. Table 5.3 demonstrates the efficacy of these calculated weights by comparing search results using them, with those obtained using a constant value of $\mu = 2$. The results are given in terms of the number of incorrect compounds listed with a better matching score than the correct spectrum. Besides yielding valuable improvements in the performance of the system studied, such work provides the basis for a more rigorous approach to the optimization of matching routines than has hitherto been available.

The Clerc system also employs bit string comparison, in this case a comparison of one bit encoded features (24a). Each of the four possible

TABLE 5.3
"Confusion" in matching 15 unknowns[a]

Compound	Constant $\mu = 2$	Calculated weights
(1) Levulinic acid	5	2
(2) Allyl acetate	2	1
(3) Phenyl isocyanate	4	2
(4) Acetamide	7	1
(5) 2-Amino pyridine	6	0
(6) 4-Methyl cyclohexanone	>10	1
(7) Benzil	>10	4
(8) Tetrahydrofuran	3	0
(9) Vanillin	>10	7
(10) Piperidine	>10	4
(11) 2-Ethyl hexyl amine	>10	4
(12) Dimethyl aniline	10	1
(13) Methyl isopropyl ketone	10	5
(14) Benzyl bromide	>10	3
(15) Allyl iodide	10	5

[a] Number of incorrect compounds listed with a better matching score than the correct spectrum. For details, see text. (Reprinted with permission from ref. 35. Copyright by the American Chemical Society.)

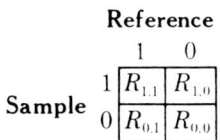

Fig. 5.7. The four weights used in the Clerc system. For details, see text. (From ref. 24a.)

results of comparison is given its own particular weight for each feature (see Fig. 5.7). For the weights R, the following relation generally holds: $R_{11} \geqslant R_{00} \geqslant R_{01} \geqslant R_{10}$. Thus, these are used to calculate a similarity rather than a dissimilarity index (see Table 5.7).

5.4.2 Comparisons Based on Intensity Ranking

In their study of computer matching of low resolution mass spectra, Knock et al. (6a) describe an extremely simple matching method as their first method.

In this, the *m/e* values of the *n* most intense peaks in reference and sample spectra are compared, irrespective of order, and the number of agreements *A* is noted. The degree of matching is then given by eqn 5.7:

$$P_1 = A/n. \tag{5.7}$$

An equivalent method was used by Law *et al.* (11a) for a small drug library, by Baty and Wade (19b) for a small steroid library and by Petterson and Ryhage (11c), as the first of the matching procedures in their tests. However, tests by Grotch (32) with this method, using 125 low resolution spectra as unknowns showed it to be significantly poorer than other methods such as a simple XOR $-\mu$ (AND) (cf. p. 121) comparison of one bit encoded full spectra.

The second method described by Knock *et al.* (6a) represents a considerable improvement. In this method, the *m/e* values of the *n* most intense peaks in reference and sample spectra are again compared, but an allowance is made for the relative positions of peaks with equal *m/e* values. The degree of matching is given by eqn 5.8:

$$P_2 = \frac{1}{n^2} \sum_{k=1}^{A} (n - |i - j|_k) \tag{5.8}$$

where *A* is the number of agreements irrespective of order and *i* and *j* are the positions in the respective sets of the *k*th pair of equal *m/e* values. An example of the use of this method is given in Fig. 5.8. This method and a further method due to Knock *et al.* (6a) in which the strongest peaks within each of *R* equal mass ranges are compared in a similar manner are currently the two methods available in the MSDC "Massmatch" system (43). Introduction of relative ranking improves the specificity and consequently the recognition performance of these systems while still maintaining a simple approach not requiring the storage of intensity data. This approach can be further improved by methods such as division into mass ranges to reduce the

Spectrum number							
(1)	m/e (I)	135 (100)	192 (54)	43 (25)	119 (23)	149 (21)	91 (15)
(2)	m/e (I)	135 (100)	43 (51)	192 (42)	119 (21)	91 (21)	51 (20)

$n = 6 \quad A = 5$

$P_2 = \frac{1}{36} [(6-0) + (6-1) + (6-1) + (6-0) + (6-1)] = \underline{0.75}$

Fig. 5.8. Comparison of two spectra of cassione (six strongest peaks) according to eqn 5.8. (Data taken from ref. 6a.)

emphasis on low mass peaks (cf. p. 108). The basis of a method equivalent to that of Knock et al. expressed in eqn 5.8 was also suggested by Tantsyrev and Kozlov (11d).

5.4.3 Reverse Search Systems using Intensity Ranking

The matching eqn 5.8 has also been used as the basis of reverse search methods by a number of authors (19a, c, d). In each case, however, these authors adopted the choice of the n most significant rather than the n most intense peaks in the reference spectrum (cf. p. 112). Sweeley and co-workers (19c) also chose one of these ions, the "designate" ion, and calculated the area under the peak due to this ion to give a quantitative estimate of the compound sought (cf. Section 8.3).

A reverse search output is shown in Fig. 5.9 (19d). In this output, in addition to the matching index, a purity index is also computed. The purity index is defined as the ratio of the fraction of the TIC accounted for by the n most significant peaks in the unknown spectrum to the fraction of the TIC accounted for by these peaks in the reference spectrum. Impure or mixed samples will give a purity index that differs appreciably from the value of one expected for a pure sample. By maintaining a small library of about 100 spectra in core, the output shown is available in real time. This real time library may also be effectively and considerably increased in size by switching another library into core at pre-set times or on detection of specified standard compounds during a GC run.

5.4.4 Comparisons made using Stored Intensity Data

One of the simplest forms of comparison using intensity data is to sum the intensity differences at all mass positions. An early reference to such a system was given by Abrahamsson et al. (11e) who calculated a disagreement index, D, where:

$$D \propto \left(\sum_{n=1}^{N} |I_{ref} - I_{unk}|_n \right). \tag{5.9}$$

This type of index was also calculated by Crawford and Morrison (5b). In fact, they used three indices such that:

$$D_1 = \sum_{n=1}^{N} |I_{ref} - I_{unk}|_n \tag{5.10}$$

SCAN	TIC	MATCH SCORE	ENTRY NO.	PURITY INDEX	COMPOUND	
320	25103					(a)
321	24435					
322	23166					
323	23548					
324	23989					
325	42530					
326	134359	0.71	78	0.98	BENZYL ACETATE (8)	
327	84199	0.71	78	0.89	BENZYL ACETATE (8)	
328	42791					
329	77120	0.71	47	0.28	MYRCENE	
330	110597	0.77	52	0.69	A-TERPINEOL	
331	49645	0.73	52	0.52	A-TERPINEOL	
332	29172					
333	34101					
334	32383					
335	26206					
336	25959					
337	26713					
338	25199					
339	24942					
340	30280					

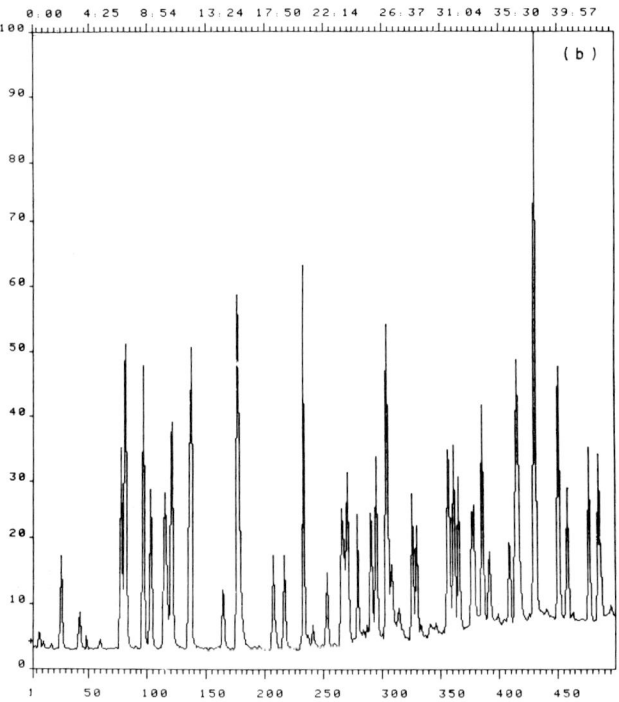

Fig. 5.9. (a) Part of real time reverse library search from perfumery mixture using small (~ 100 spectra) core-based library. Components identified are benzyl acetate and α-terpineol. Notice low purity index for initial misidentification of α-terpineol as myrcene. (b) TIC record from same mixture. (Courtesy Kratos-AEI Scientific Instruments).

for spectral intensities normalized so that:

$$\sum_{n=1}^{N} I_n = 1 \qquad (5.11)$$

$$D_2 = \sum_{n=1}^{N} |I_{ref}^{\frac{1}{2}} - I_{unk}^{\frac{1}{2}}|_n \qquad (5.12)$$

for spectra normalized so that:

$$\sum_{n=1}^{N} I_n^{\frac{1}{2}} = 1 \qquad (5.13)$$

and

$$D_3 = \sum_{n=1}^{N} (I_{ref} - I_{unk})_n^2 \qquad (5.14)$$

for spectra normalized so that:

$$\sum_{n=1}^{N} I_n^2 = 1. \qquad (5.15)$$

All three comparison equations gave satisfactory results, although it was found that normalization by eqn 5.13 was to be preferred when scatter was added to the peak intensity data. Similar indices were used by Grotch (15a) with spectra abbreviated to the mass of the most intense peak in each 14 a.m.u. interval together with two bits of intensity information. The two indices used were:

$$D_4 = \frac{\Sigma (I_{unk} - I_{ref})_n^2}{\Sigma (I_{unk}^2 - I_{ref}^2)_n}, \qquad (5.16)$$

and

$$D_5 = \frac{\Sigma |I_{unk} - I_{ref}|_n}{\Sigma (I_{unk} + I_{ref})_n}. \qquad (5.17)$$

Equation 5.16 is a normalized least squares fit and eqn 5.17 is a normalized sum of the absolute difference criterion. For a perfect match both criteria become zero. The eqn 5.17 yielded slightly better results than the least squares criterion (5.16). Surprisingly, the identification performance using eqn 5.17 with abbreviated spectra was better than that found using full spectra although the average number of bits used was about two orders of magnitude less. This may result from a smoothing out of data variability as a result of the coding method used. Further references to the use of similar intensity based disagreement indexes are available (11b, 36).

Rather than use absolute intensity differences as a matching criterion, some authors have preferred to use a relative measure of change at each matching mass in the reference and the unknown. Examples of this method are the use of the intensity difference as a fraction of the intensities used to generate the difference or the simple intensity ratio between reference and unknown. From practical considerations there seem to be good reasons for preferring the use of a relative measure. Variations in spectra belonging to the same compound, which may, for example, be due to mass discrimination or statistical effects, will generally take the form of a fractional rather than an absolute variation in ion intensity. Thus, an intensity of 2% in the reference and 20% in the unknown would be held to indicate far more dissimilarity than 32% in the reference and 50% in the unknown, although the absolute differences are the same.

An idea of the practical application of intensity ratios in a matching system may be gained by a study of the method used by Biemann et al. (13a). In this, the ratios of intensities of peaks at the same mass in the known and unknown spectra (both abbreviated to two masses every 14 a.m.u.) are calculated, starting at the lowest mass common to both full spectra. If a given mass is present in only one spectrum, the corresponding ratio is set to zero. Ratios much lower or much higher than unity may result either from great differences in mass discrimination (caused by different instruments or smooth changes in sample concentration during the emergence of a gas chromatographic peak) or from actual non-identity of the spectra. To distinguish between these two cases, all ratios are divided by the average of those ratios due to large peaks ($>10\%$ average intensity). In the case of actual non-identity, the average ratio of large peaks is far from unity and division by this value generates even greater differences in the ratios. For similar spectra however, the average ratio is near unity and division but this value causes little change in the ratios. Finally, any ratios which are greater than unity are replaced by their reciprocals, the ratios weighted to increase the contribution of ratios calculated from more intense peaks and the average weighted ratio calculated.

Biemann uses a further matching factor, that is the fraction of the total ionization from both spectra due to peaks that do not have corresponding masses in the other spectrum. This "fraction of unmatched intensities" is used to calculate the similarity index (SI):

$$SI = \frac{\text{average weighted ratio}}{\text{fraction of unmatched intensities} + 1}. \quad (5.18)$$

This SI has a theoretical maximum of 1·00. McGuire et al. (6b) have suggested that an SI value $<0·2$ is a poor match, a value of $0·2–0·35$ represents a match with a related compound and that a value $>0·35$ is a good match if the SI for the second best compound is significantly lower (cf. Fig. 5.15). A detailed discussion of the implementation of the Biemann algorithm has been given by Hoyland and Neher (37).

Bell (17) has adopted a method based on that of Biemann et al. using three peaks every 14 a.m.u. and a slightly different approach to the problem of mass discrimination. In this case, the ratios of the intensity of the largest peak to that of the second largest and of the largest to the third largest are calculated both for the library spectrum and the unknown (L_1, L_2 and U_1, U_2 respectively in Fig. 5.10). The ratio of ratios is calculated for each set of matching peaks, thus $R_1 = L_1/U_1$ and $R_2 = L_2/U_2$. Any ratios greater than unity are replaced by their reciprocals and these corrected ratios averaged to give a SI. A method similar to that of Biemann has been used by DeGragnano (15b).

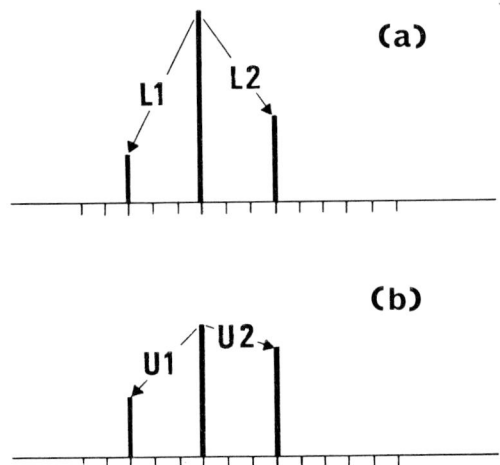

Fig. 5.10. Calculation of the abundance ratios for the three largest peaks in each 14 a.m.u. region. (a) Library spectrum. (b) Unknown spectrum. (From ref. 17.)

The MSSS of Heller (2, 38) makes use of intensity data in its major search routine, the "peak and intensity" search. MSSS is an interactive, conversational mass spectral retireval system, available over ordinary telephone lines from the CYPHERNET time-sharing computer network. In the peak and intensity search, the user specifies a significant mass present in the unknown together with its intensity. As previously noted (p. 102), MSSS uses an inverted file for this search, i.e. the mass spectral data is indexed on the presence or absence of a peak at each mass number in an abbreviated spectrum. Thus, on specification of a mass, e.g. m/e 85, the corresponding (85th) cell in a pointer file points to the location of all references to that mass and their associated intensity values. The intensity check uses a range factor, in this case two, so that an intensity of 20% will be interpreted as a range of intensities from 10%–40%.

The program responds to the user's suggestion of a mass and intensity with the number of spectra in the file that have a peak at that mass within the required intensity limits. After the user has entered the second mass and intensity, the program responds with only those spectra that contain the required combination of peaks and intensities. At any stage, when the number of spectra remaining is sufficiently small, the details of these spectra may be listed (see Fig. 5.11).

The mass spectral data in MSSS is indexed in a number of ways other than by peak presence so that a wide choice of search routines is available. A

```
PEAK AND INTENSITY SEARCH

USER: INTENSITY RANGE FACTOR IS: 2

TYPE PEAK,INT
CR TO EXIT, 1 FOR ID,MW,MF AND NAME

USER: 85,100

    # REFS      M/E PEAKS

     214           85

NEXT REQUEST: 128,20

    # REFS      M/E PEAKS

      11          85  128

NEXT REQUEST: 29,30

    # REFS      M/E PEAKS

       8         85  128  29

NEXT REQUEST: 1

   ID#      MW       MF                      NAME
   1671     170   C10.H18.O2     GAMMA–DECALACTONE
   1688     254   C16.H30.O2     GAMMA–PALMITOLACTONE
   1689     282   C18.H34.O2     GAMMA–STEAROLACTONE
   6731     128   C8.H16.O       4–OCTANONE
   6736     128   C8.H16.O       2–METHYL–3–HEPTANONE
   6740     128   C8.H16.O       2–METHYL–4–HEPTANONE
   6747     128   C8.H16.O       2,5–DIMETHYL–3–HEXANONE
   8699     314   C19.H38.O3     METHYL 4–HYDROXYOCTADECANOATE
```

Fig. 5.11. Mass Spectral Search System (MSSS) "peak and intensity" search. Conversation and output. (From ref. 28.)

recent list of routines includes peak and intensity, loss and intensity, mol. wt, molecular formula and MSDC compound classification code searches as well as a number of combinations of these options (18c). Properties such as mol. wt and molecular formula and the MSDC code can be an effective filter in narrowing the list of compounds retrieved by a procedure such as the peak and intensity search.

MSSS does not use a similarity index as an integral part of the search. It has been suggested that, since the search itself gives rise to so few answers, the user himself can perform a "similarity index" match by comparing the full file spectra with the unknown spectrum (38). A dissimilarity index (eqn 5.19) is available as an option and has been designed to emphasize differences between spectra by giving extra weighting in cases where there is a peak in one spectrum but not in the other.

Where
$$\text{Index} = \left[\sum_{n=12}^{400} |I_{unk} - I_{ref}|_n^k /400 \right]^{\frac{1}{2}} \quad (5.19)$$

$k = 2$ if I_{unk} or $I_{ref} = 0$, otherwise $k = 1$.

5.4.5 Substructure Searching using Stored Intensity Data

The Self-training Interpretative and Retrieval System (STIRS) described by McLafferty and co-workers (23) also incorporates match factors based on the comparison of intensity data (see p. 114). However, in STIRS, separate classes of data (e.g. characteristic ions, ion series and neutral losses) are identified and for each class the appropriate data for the unknown is matched against the library data to give a match factor (MF) for that class. As an example, the three most abundant odd mass ions and three most abundant even mass ions below m/e 90 are used to calculate separately MF_{odd} and MF_{even} from eqn 5.20 where I_i and I_j are the intensities of the two peaks in the kth match, n is the number of odd or even mass characteristic ions in each spectrum and A is the number of agreements:

$$\text{MF} = 1000 \left[\sum_{k=1}^{A} (I_i/I_j)_k \right] \bigg/ \left[\sum_{k=1}^{n} (I_i + I_j)_k \right]. \quad (5.20)$$

Match factor 2 (MF2) in STIRS is then the mean of MF_{odd} and MF_{even}. The output for each class comprises the compounds with the highest match factors for that class.

The major application of STIRS has been as an aid to the interpretation of spectra to indicate the presence of particular substructures in the unknown. In the original studies, for example, a structural feature in the unknown was indicated by MF2 when the substructure was present in at least five different compounds in the top 10 list which had MF2 > 500. This particular match factor can indicate the presence of a wide range of features from a simple ether linkage to much more complex structures.

More recently, the compounds selected by STIRS have been examined, using the computer, for the presence of particular substructures to provide a statistical evaluation of the probability of the presence of that substructure in the unknown (27). For a set of 188 compounds, STIRS selected the 15 compounds with highest MF values (excluding the examined compound itself) for each of the 11 data classes. The average match factor (AMF) for the kth data class and the lth substructure, $AMF_{k,l}$ was calculated from eqn 5.21 where $MF_{k,l}(i)$ is the MF_k value of the ith compound of the n selected compounds containing substructure l ($n \leq 15$). A distribution of AMF values for the phenyl group together with the AMF values obtained when phenyl was not present in the original compound, is plotted in Fig. 5.12. A preliminary examination of such distributions shows the cases in which the "present" and "not present" distributions do not exhibit a useful separation and these combinations of k and l are not used further:

$$AMF_{k,l} = \sum_{i=1}^{n} MF_{k,l}(i)/15. \qquad (5.21)$$

In the analysis of unknowns, each AMF value is converted, using the mean and standard deviation of the established distributions (see Fig. 5.12), to the probability that substructure l is present ($P+$). In addition, a knowledge of the reference library composition is used to establish the probability that the chosen compounds could have resulted from random selection ($P-$). Then the confidence level is given by eqn 5.22:

$$\% \text{ confidence} = \frac{(P+) \times 100}{(P+) + (P-)}. \qquad (5.22)$$

For example, from Fig. 5.12 an $AMF_{11,R}$ (MF 11, phenyl group) value of 225 for an unknown spectrum should correctly indicate the presence of a phenyl group in 80 out of 100 cases.

The published data of McLafferty and co-workers includes a comprehensive table (see Table 5.4) giving the recall ability of individual MFs for 22 functional groups at the 80% confidence level. In addition to providing confidence levels to assess the results obtained with unknowns, this work provides a basis for a more systematic assessment of new and improved data classes for substructure analysis (see p. 115).

Extensions of STIRS to the prediction of data other than the presence of substructures have been made (39). For example, the average double bond equivalent (DBE) of the 15 compounds selected by STIRS with the highest overall match factor from a data set restricted to the common elements

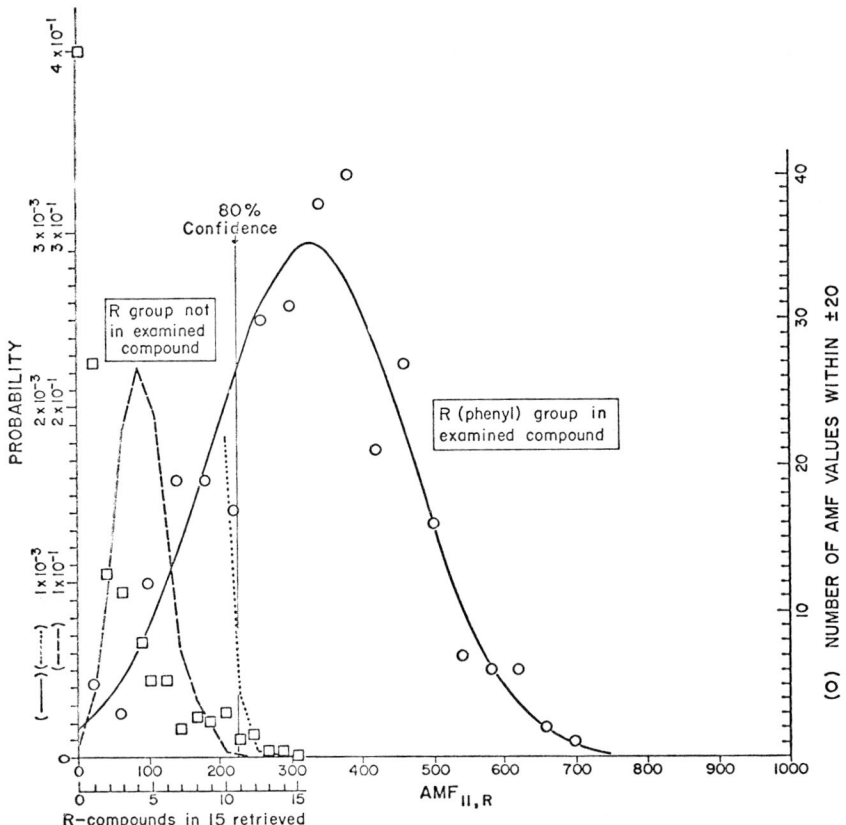

Fig. 5.12. Distribution of average match factor values $AMF_{11,R}$ for data class MF_{11} (overall match factor) and functional group R (phenyl), assuming a Gaussian distribution (solid line), based on 300 examined spectra of compounds selected at random whose WLN structure codes contain the symbol R. The experimental values (circles, referred to right hand ordinates) represent the number of spectra found with $AMF_{11,R}$ values within ± 20 units of the abscissa value shown. The dashed line (abscissa "R-compounds in 15 retrieved") represents the probability of randomly selecting a particular number (j) of R-containing compounds out of 15 from the same file which contains 28.6% such compounds. For the dotted line, these probability values are shown at $100 \times$ sensitivity, the same scale as used for the $AMF_{11,R}$ values above. The abscissa is converted from j values to $AMF_{11,R}$ values by setting the $j = 15$ value equal to the mean of all values of the examined compounds not containing R, $MF_{11,-R}$. The squares represent the proportion of examined compounds not containing phenyl out of 300 total (ordinate) for which a particular number of the 15 selected compounds (abscissa) do contain phenyl. In this case, the proportion is less than predicted by the random selection model; indication that STIRS discriminates against the selection of phenyl. (From ref. 27.)

TABLE 5.4

Recall ability (%) of STIRS at 80% Confidence Level[a]

WLN	Functional group[a]	% in file	MF1[b] Ion series m/e 27–99	MF2 27–89	MF3 90–149	MF4 ≥150	MF5 0–64	MF6 ≥65	MF7[c] I I°	MF10 2 peaks/ 14 a.m.u.	MF11 Overall, ΣMF1–6
1	—CH$_2$—, CH$_3$	47.2		13	13		26		6	14	49
2	—CH$_2$CH$_2$—, C$_2$H$_5$	20.2		12			19		6		18
≥3	Alkyl ≥ C$_3$	20.3	22	40	32	19	38			43	64
V	Carbonyl	46.2	16	18	19	30	39	18		20	56
Q	Hydroxyl	20.3	21	8		20					40
O	Linking O	41.0		35	32		49	5		41	62
VO	Ester, anhydride	19.4		14	11		41	7		30	63
R	Phenyl	28.6	41	67	54	20	28	28		64	72
Z	I° amine	4.3				7					21
M	II° amine	11.9									
N	III° amine	28.8	1	34	8		3			28	48
S	Sulphur	10.4		39	10		14			29	53
O1	Methoxy	19.0		37	10					27	40
O2	Ethoxy	5.2								41	21
L...J	Carbocyclic	24.1		52	46		21	3	6	40	56
T...J	Heterocyclic	52.3	48	61	58		23	6	7	75	76
U	Double and triple bonds	23.9		24	20		5		1	12	36
Y	Single branch	26.4	4	26	16		12		3	15	36
X	Double branch	10.9		44	33	5				19	35
G	Chlorine	8.2		34	47		71		41	29	74
E	Bromine	3.4					44			22	22
F	Fluorine	5.0		27			33				62

[a] From ref. 27.

C, H, N, P, O, S, Si, F, Cl, Br and I in their natural isotopic abundances provided a useful prediction of the DBE for an unknown. Thus, 87% of predictions were within ± 1 for true DBE values of 0–4 and approx. 80% within ± 2 for true DBE values of 5–11.

5.4.6 Reverse Search Systems using Stored Intensity Data

A number of reverse search systems using intensity data have been described. The system of Abramson (1) using library spectra consisting of the 10 most intense peaks, begins by normalizing the intensities in the unknown spectrum at these masses using the library base mass as base mass for the unknown. Comparisons are then made by calculating the intensity difference between unknown and library spectra as a fraction of the smallest intensity used to generate the difference. If, during a comparison, a relatively large number of large negative fractional differences are found, renormalization to the second largest peak is attempted on the basis that interference at the base peak in the unknown has distorted the relative intensities. Comparison is then repeated. Positive deviations also arise because of interferences and a certain number of these are allowed. Thus, the search can be made relatively immune to interference. As an example, Abramson (1) quotes the detection of the drug methaqualone in an artificial background of perfluorokerosene (PFK) and cyclohexane added so that the base peak of methaqualone was only the ninth largest peak in the combined spectrum.

The average fractional difference is used as the basis for a matching index. Providing that this average fractional difference does not exceed $\pm 1/6$, a match is declared, and the intensities of the masses searched for in that scan are summed. These summed intensities are used as the basis for a quantitative estimation of the compound sought. In addition, the name of the compounds sought and the best match index obtained from these scans are printed out. An alternative search system developed by Grönneberg et al. (19a) which searches for the N most significant peaks (see p. 112) performs a similar function in that it returns as a score the sum of the relative intensities in the unknown of the significant ions, characterizing the reference compound that was sought after. An additional requirement in this system is that all these significant ions must be found in the unknown spectrum.

Another application of the reverse search principle has been devised and used by McLafferty and co-workers (40). In this "probability based matching" (PBM), unknown spectra are, as usual, matched to the abbreviated spectrum of the single compound sought (referred to as the "target compound"). The basis of the match is a confidence index, K, derived such that 2^K represents the number of compounds, selected at random, whose mass spectra would have to be examined to find data which matches the target spectrum to the same degree as does the unknown. Thus, the greater the value of K, the more certain the match:

$$K = \Sigma K_j = \Sigma(U_j - A_j - D + W_j). \tag{5.23}$$

The K value (eqn 5.23) is the sum of individual values, K_j, for each peak in the abbreviated target spectrum (usually 10–15 peaks) whose abundance falls within required limits in the unknown spectrum. K_j is assumed to be a linear combination of four independent probabilities. U_j is the uniqueness of the m/e value of the jth peak; A_j is a modification of U_j based on the abundance of the peak in the target spectrum; D is a dilution factor to allow for dilution of the target compound by other materials in the unknown and W_j is a window tolerance used in comparing the relative abundance in the unknown with that in the target spectrum. A summary of the principles used in deriving these probabilities is given here, but for further details, the reader is referred to the original paper (40).

The U value is based on the probability that the abundance of a peak at the mass in question would be $> 50\%$ of the base intensity in a spectrum taken at random. U values were initially determined from the abbreviated spectra of 17 124 compounds listed in the "Eight Peak Index of Mass Spectra". A mass such as m/e 43, which is very common, has a low value ($U = 3$) whereas high mass peaks which are more diagnostic have higher values, e.g. $U = 10$ for all masses above m/e 300. For higher masses, the reference file could be poorly representative and thus average U values are assigned to mass ranges and a maximum value of $U = 10$ recommended for magnetic sector machines. Corrections are also made to U values to allow for the use of a quadrupole MS in the associated experimental work.

The probability of occurrence of a peak at a particular mass increases rapidly with decreasing intensity. Thus, the U value must be reduced to evaluate the uniqueness of peaks of less than 50% intensity. Experimental results obtained by both McLafferty (40) and Grotch (32) indicate a lognormal distribution of intensities in spectra, i.e. an equal number of peaks in ranges covering the same factor increase in intensity. Thus, McLafferty proposed the abundance values (A) listed in Table 5.5 to reduce the uniqueness value U for lower intensities, the intensity value being taken from the target spectrum. A few masses showing different intensity distributions were accommodated by using a different initial U value.

A later, more extensive, investigation into the statistical occurrence of mass and abundance values in mass spectra used the Wiley Registry containing 18 806 full spectra to determine uniqueness values based on the probability of occurrence of peaks of $\geq 1\%$ relative abundance at each mass (41). So called "U_0" values calculated from a data base of compounds of mol. wt higher than the mass of interest were considered to be most useful results and were found, above m/e 115, to show a nearly linear relationship with mass, viz: $U_0(m) = 0.65 + 0.0077m$. Abundance values obtained from this investigation also supported the log-normal abundance distribution established previously.

TABLE 5.5

Effect of abundance on peak uniqueness[a]

"A" Value	Abundance range	
	PBM	Grotch[a]
0	50–100%	42–100%
1	19–50	22–42
2	7·1–19	12–22
3	2·7–7·1	4–12
4	1·0–2·7	1·0–4
5	0·38–1·0	

[a] Data from refs 32 and 40.

TABLE 5.6

Effect of window tolerance on peak uniqueness[a]

Abundance range of window tolerance	"W" Value
10·4%	5
20%	4
37%	3
63%	2

[a] From ref. 40.

The data in Table 5.5 is also used to calculate a reduction in K (equal to D) when the target compound is diluted by other materials in the sample. For example, if the base peak in the target spectrum has only 15% intensity in the unknown, a value of 2 is subtracted from the K_j value for each peak used. The relative abundances of the peaks in the unknown spectrum have to be consistent with those of the corresponding peaks in the target spectrum for a match to be found. The closer the unknown intensities are to those of the target compound, the less the chance that this could have occurred by random selection and the greater the W value added to U (see Table 5.6).

For a window tolerance of 20% (with this tolerance, a peak ratio of 0·010 would include ratios in the range 0·008–0·0125), there are 16 possible intensity windows in the normal intensity range, i.e. 0·1%–100%. Thus, there is only one chance in 2^4 that a peak selected at random will have an intensity value within the 20% window, so that fulfilment of this criterion adds 4 to the K value. If the peak intensity falls outside this tolerance, $K_j = 0$. Naturally, one peak in the reference spectrum must serve as the intensity reference, so for this peak $W = 0$. In practice, window values of 20% or 37% were used.

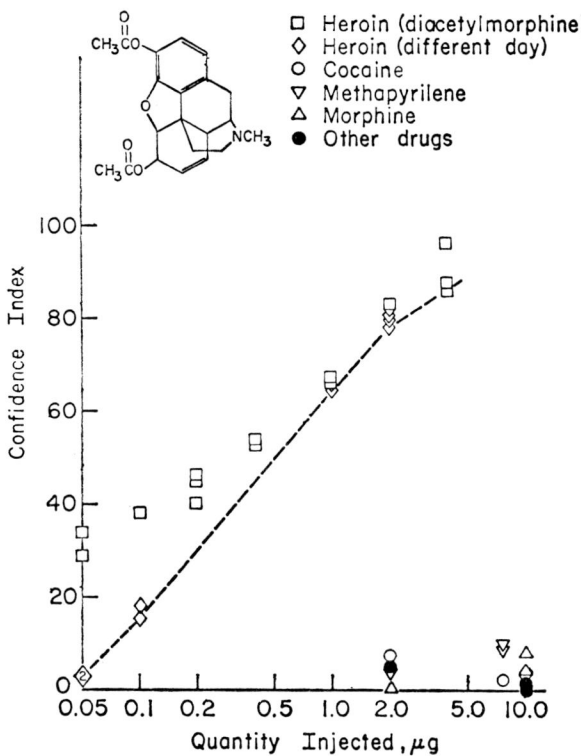

Fig. 5.13. Heroin identification algorithm: effect on Confidence Index of the amount of heroin and of other drugs as challenges. Other drugs tested include quinine sulphate, procaine hydrochloride, ephedrine, strychnine sulphate, amphetamine sulphate, metamphetamine hydrochloride, secobarbital, phenobarbital, amobarbital and pentobarbital. Small numbers within the symbols indicate the number of determinations yielding the same value. (From ref. 40.)

The PBM system was extensively tested by the authors on a variety of compounds and typical data is shown in Fig. 5.13. Hardware identification programs on permanent cards based on PBM principles, and unique for each compound were later used as an important part of an automatic quadrupole MS system designed to achieve real time identification of specific compounds in complex samples (42). Subsequent applications of PBM used a procedure modified to permit the presence of a limited number of peaks in the unknown spectrum of lower relative intensity than in the reference spectrum. Sensitivity of the method to impurities and other errors in the reference spectrum was decreased in this way and the recall of the system considerably improved (19e).

5.5 FILTERING

The filter or pre-search takes the form of a relatively simple test that can be carried out much more quickly than the full match and which will eliminate a majority of library spectra as being too dissimilar to the unknown to be worth complete matching. The object of the pre-search is, of course, to reduce the time needed to compare an unknown with the library.

Probably the most common form of pre-search is to establish the coincidence in mass of the major peaks in the unknown and library spectra. This test has been used with varying degrees of severity. For example, Crawford and Morrison (5b) required that two or three of the largest peaks in both the unknown and reference, coincided without regard to order, whereas a filter available with the Massmatch system (43) requires only that the base peak of the unknown should be amongst the six most intense in the library spectrum. The system of Biemann (13a) requires that the unknown base peak shall be $>25\%$ in the reference spectrum and that the reference base peak is $>25\%$ in the unknown spectrum. A system described by Grotch (32) has a very similar requirement, viz. the unknown base peak should be amongst the five highest in the reference spectrum and vice versa.

Another form of pre-search is the specification of a mol. wt range for the reference spectra to be matched (5b, 6a, 11f, 16, 43). The system of Robertson and Merritt (see p. 109) (16) incorporates an ingeniously coded form of mol. wt. range filter. In their coding system, each computer word stores information which covers a mass range of 35 a.m.u. As many words as are needed are used to encode the spectrum and the last word is designated by setting a flag in the otherwise vacant 16th bit. During the search, only those reference spectra with the same number of words as the unknown are searched and, in fact, the library is divided into sub-libraries based on the numbers of words used. Note should also be made of the MSSS (see p. 129) which, in effect, consists of a series of cleverly designed filters for peak presence, molecular formula, mol. wt etc.

A pre-search described by Biemann (13a) also eliminates compounds that differ greatly in mol. wt. Thus, the number of peaks in the abbreviated reference spectrum must agree with the number in the abbreviated unknown spectrum $\pm 75\%$. In addition, the Biemann system employs another filter based on a comparison of the five largest members of the 14-element ion series obtained by summing the intensities of the ions at m/e $1 + 14n$, $2 + 14n, \ldots, 14 + 14n$, (where $n = 0, 1, 2, \ldots$) in both the reference and unknown spectra. Other filters based on equivalence of carbon number or GC retention time have also been proposed (43, 44) (cf. p. 142).

Another ion series related filter, the series displacement index (SDI) has been used by Dromey (45).

$$\text{SDI} = \sum_m d_m \times I_m. \qquad (5.24)$$

The index is calculated from eqn 5.24 where d_m is the positive displacement (in a.m.u.) of the mth ion in the abbreviated spectrum from the closest smaller alkene series mass (i.e. 41, 55, 69, 83, ...) and I_m is the intensity of the mth ion normalized relative to the TIC for the spectrum. In practice, the file of library spectra is ordered according to series displacement indices so that only a small region of the library, dictated by the SDI value for the unknown, is needed for the search. A search range of ± 50 SDI was found to be sufficient to allow for variations between duplicate spectra whilst requiring, on average, a file length of only 22% of the full library. This file length can be further restricted by a factor of one and half to two by the use of a mass centroid index calculated from eqn 5.25:

$$M_c = \Sigma M_j I_j \qquad (5.25)$$

where M_j is the jth mass, I_j is the intensity of the jth mass and $\Sigma I_j = 1$.

As can be seen from Fig. 5.14, members of a particular chemical class tend to cluster in a narrow region of the table. Thus, a search restricted by SDI should still retrieve spectra belonging to a particular class that are very similar to the unknown.

The advantages of a pre-search are improved search speeds and the possibility of less ambiguous identifications by filtering candidate spectra. To set against this, the main disadvantage of a pre-search is that the true compound and, even more likely, related compounds may be excluded from consideration if the pre-search conditions are too stringent. For example, if the operator is reasonably certain of the mol. wt of the unknown, a mol. wt restriction will bring obvious dividends in speed and reliability of identification, so long as the unknown is in the library, whereas if the unknown is not in the library, compounds of similar structure that fall outside the mol. wt range will not be retrieved.

A better filtering system, in cases where it can be applied, is the threshold or hurdle system described by Clerc (24). Using the binary coding system

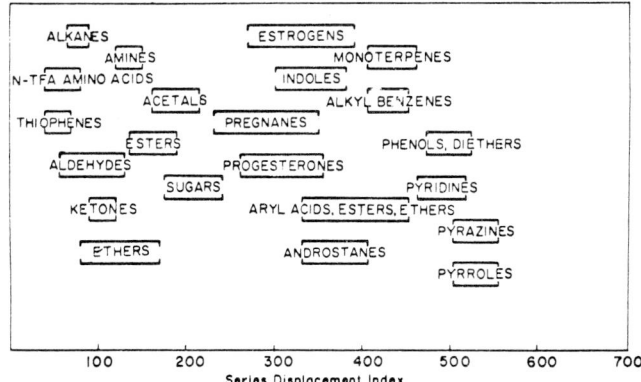

Fig. 5.14. Table showing the Series Displacement Index (SDI) ranges occupied by different classes of organic molecules. More than 10 spectra were used in the results for each class. The delimiters above and below the class names show the SDI ranges. (Reprinted with permission from ref. 45. Copyright by the American Chemical Society.)

previously described (p. 114), the relative ability of each feature encoded from the unknown spectrum to discriminate against dissimilar spectra is computed before matching. A measure of this discriminating ability is given by the difference $\Delta R = R_{11} - R_{10}$, if the corresponding feature in the unknown is encoded as a 1 and by the difference $\Delta R = R_{00} - R_{01}$, if it is encoded as a 0 (p. 123). These calculations are carried out without reference to library spectra. The values of ΔR are then used to determine a priority sequence for feature comparison, the features with the largest ΔR values being compared first. Thus, it is possible to define a threshold value for the similarity index which must be exceeded after comparison of a very small number of features, otherwise the whole comparison is aborted. The important point is that this priority sequence is determined solely from the signature of the unknown, so that the test is hopefully optimal for each unknown. Savings of 70% of the comparison time may be obtained by this method.

The proviso made regarding the application of a pre-search does not apply so rigidly in the case of a reverse search. For example, in a reverse search for a particular compound, the base peak of this compound must of necessity appear with a reasonable intensity in the unknown spectrum. Thus, the pre-search condition used by Abramson (1) in his reverse search in which at least five of the 10 masses sought must have net (i.e. scan minus background) intensities greater than a set threshold, is quite justified.

Again, in the probability based reverse search of McLafferty (40) two modifications are used in the autoscan mode which greatly increase the speed of operation. Masses are examined in a priority sequence based on diminishing values of $(U_j - A_j)$ (see p. 136) and this examination is terminated when the sum of the K_j values determined exceeds a pre-set threshold, K_T, at which the identification is assumed to be sufficiently positive. Further, a negative answer is given, i.e. the compound sought is absent, as soon as the product $I_j S_j$ (I_j is the intensity at the *j*th mass and S_j is the sensitivity of the *j*th mass for the compound sought) for the mass examined indicates a quantity below a present minimum value, Q_T.

Another example of a pre-search which can only with safety be applied in a reverse search is the limitation of a search of GCMS data by use of GC retention times. However, in a conventional forward search the matches obtained may be subsequently interpreted in the light of GC retention data and, indeed, any other relevant data. An example of this type of application may be seen in Fig. 5.15 which lists some library search results obtained during the analysis of the urine extract from a comatose patient (44b). With the barely resolved GC peak at scan 161, the search routine only finds matches with very low similarity indices, so that no definite conclusion as to the identity of the material is possible. However, the fact that "propoxyphene metabolite" has the same retention index within three units while others differ by $+184$ and -71 units, respectively (the limit of accuracy being ± 5 units (44a)), makes it practically certain that the former is the correct identification.

The matching scores for scans 169 and 175 are higher and thus there is less need for the use of confirmatory retention index data. However, it can be seen that in a reverse search situation, the use of a retention index window or filter for the search would have considerably increased the speed of comparison and the reliability of identification, since components of the mixture which elute outside this retention index window do not interfere, even if they have similar mass spectra. The most interesting example is the search result for scan 204. Although none of the compounds listed can be identical because their retention indices are so different, it is reasonable to conclude from the first two choices that the unknown may be related to chlorpromazine. It had previously been noted (44a) that aromatic oxygenation increases the retention index by about 550 units and it was therefore possible that this compound could be similarly oxygenated. In fact, further comparison of the unknown spectrum with standard spectra established the identity of the compound as chlorpromazine *S*-oxide.

A prerequisite for the efficient use of retention index data in this way is the automatic generation of this parameter together with the mass spectra. This facility is provided by a system for computer assisted assignment of retention indices in GCMS described in a paper by Nau and Biemann (44a). This paper also reports the application of the system to the identification of drugs

```
              SCAN = 161      I(X) = 2751

              SEARCH RESULTS              SI       I(A)   I(A)-I(X)
PROPOXYPHENE METABOLITE 1                0.138    2748    -    3
PROPOXYPHENE METABOLITE 3                0.134    2935    +  184
DEALKYL CHLORPROMAZINE                   0.130    2680    -   71
NORTRIPTYLENE                            0.073    2565    -  186
INDOMETHACIN                             0.068    3192    +  441
2,6-DI-T-BUTYL-4-METHYL PHENOL           0.067    1630    -1121


              SCAN = 169      I(X) = 2864

              SEARCH RESULTS              SI       I(A)   I(A)-I(X)
CHLORPROMAZINE                           0.679    2855    -    9
DESMETHYL CHLORPROMAZINE                 0.232    2940    +   76
PROCHLORPERAZINE                         0.153    3468    +  604
DEXBROMPHENIRAMINE                       0.087    2414    -  450
LIDOCAINE                                0.068    2174    -  690
PERPHENAZINE                             0.065    4015    +1151


              SCAN = 175      I(X) = 2952

              SEARCH RESULTS              SI       I(A)   I(A)-I(X)
DESMETHYL CHLORPROMAZINE                 0.248    2940    -   12
CHLORPROMAZINE                           0.116    2855    -   97
METHAQUALONE METABOLITE 2                0.087    3105    +  153
METHAQUALONE METABOLITE 1                0.080    2905    -   47
PROCHLORPERAZINE                         0.078    3468    +  516
LIDOCAINE                                0.063    2174    -  778


              SCAN = 204      I(X) = 3408

              SEARCH RESULTS              SI       I(A)   I(A)-I(X)
CHLORPROMAZINE                           0.197    2855    -  553
DESMETHYL CHLORPROMAZINE                 0.148    2940    -  468
PROCHLORPERAZINE                         0.066    3468    +   60
TRIFLUPROMAZINE                          0.066    3038    -  370
DEALKYL CHLORPROMAZINE                   0.065    2682    -  726
NORTRIPTYLENE                            0.065    2565    -  843
```

Fig. 5.15. Search results using a library of 300 mass spectra of known drugs, metabolites and frequent constituents of body fluids. Column 2, SI is similarity index. Column 3, I(A), retention indices of the search finds. Column 4, I(A)–I(X), difference between I(A) and the retention index of the scan corresponding to the "unknown". (Reprinted from ref. 44b, by courtesy of Marcel Dekker Inc.)

in body fluids and to the complete characterization of the amino acids and oligopeptides from the hydrolysis of derivitized polypeptides (see p. 193). A similar system for the assignment of retention indices has been described by

Sweeley and coworkers who have also demonstrated the use of this data to limit the number of spectra to be compared and to increase the reliability of identification in a reverse search system (19c).

5.6 METHODS RELATED TO LIBRARY SEARCHING

Jellum and coworkers (46) have used an extension of the library search principle as the basis of a method for the recognition of anomalous compounds in multicomponent mixtures Computer Assisted Search for Anomalous Compounds (CASAC). Repetitively scanned spectra from a GCMS analysis of the mixture are compared with a library which consists of a file of spectra from an identical analysis of a "normal" mixture. The comparison procedure allows for variation in GC retention times by matching each unknown against library spectra within a window of about 20 scans. A simple eqn 5.26 or 5.27 is used to determine the degree of coincidence, C, and this value is plotted against scan number:

$$C^* = \frac{N_{LS}^2}{N_S N_L} \cdot 1000 \text{ (for } N_S < N_L) \qquad (5.26)$$

$$C^* = 2000 - \frac{N_{LS}^2}{N_S N_L} \cdot 1000 \text{ (for } N_S > N_L) \qquad (5.27)$$

where N_{LS} = number of common peaks in the defined mass range, N_L = number of peaks in the library spectrum and N_S = number of peaks in the sample spectrum. The two methods of calculation for C^* are employed to distinguish between the cases of missing and additional compounds in the unknown mixture. The method has been used to locate abnormal compounds in urine and an example of this application is given in Fig. 5.16.

Very promising results have been obtained using this method with data from high resolution capillary column analyses (47) and it is anticipated that the method could profitably be applied in other fields, e.g. quality control.

5.7 SUMMARY

Library search methods represent one possible approach to the identification of unknown compounds. If an exact match is not possible, a library search that has not been unduly restricted by pre-search conditions forms a very useful first step in the identification routine in providing suggestions towards the structure of the unknown. At this stage, the output of the search may be interpreted by the user in the light of other analytical data that may

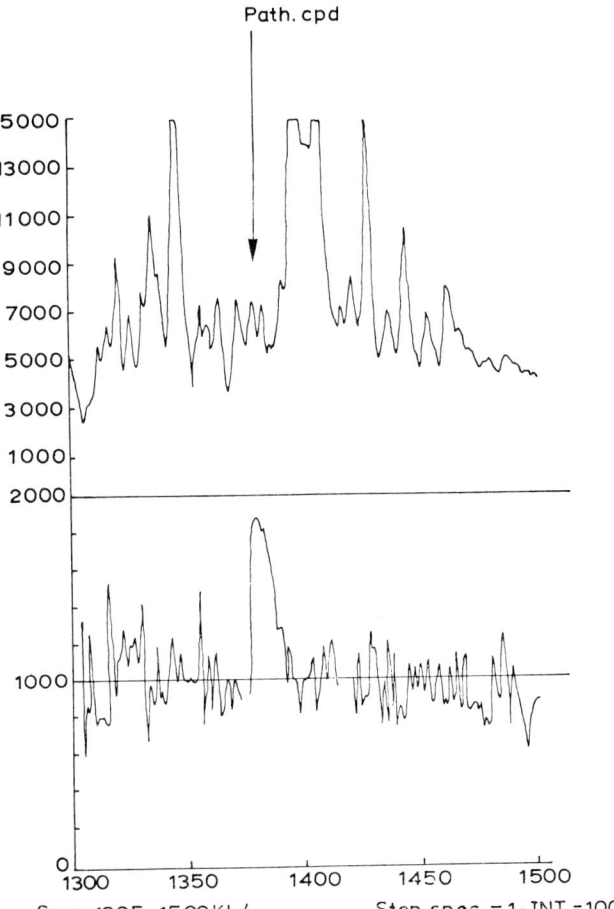

Fig. 5.16. Localization of an abnormal compound by the CASAC program. A library was generated from the normal, pooled urine sample (organic acid methyl esters), which was separated on an OV-17 column using temperature programming from 80°–210°. About 200 mass spectra were recorded and converted into a library. A urine sample from a patient was treated in exactly the same way, and automatically matched against the library. Scan number 1375–1385 showed poor coincidence (10–20%) because of the presence of an abnormal compound, which was identified by an off-line library search as a cyclopropanedicarboxylic acid. (From ref. 46.)

be available. Examples of this procedure are the subsequent use of elemental compositions from accurate mass data, particularly when this data can be obtained at low resolving power (Table 5.7), or of GC retention data (see p. 142). Manual comparison of full spectra is also accepted as necessary to confirm matching results in most cases (48).

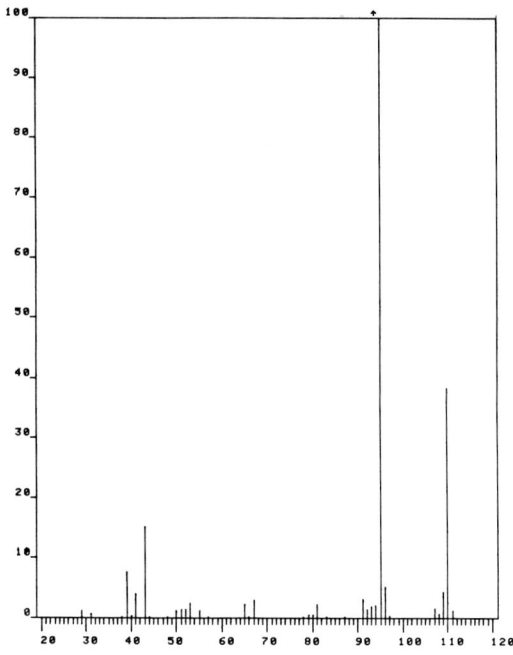

Fig. 5.17. Low resolution scan corresponding to library search output in Table 5.7.

TABLE 5.7

Summary of library search and mass measurement data from a single low resolution scan of an unknown (Fig. 5.17) [a]

Formula	% Score	Compound
$C_{12}H_{20}O$	96·5	2,5-di-tert-butylfuran
$C_6H_6O_2$	95·3	2-furyl methyl ketone
$C_6H_6O_2$	95·3	furyl methyl ketone
$C_7H_8O_2$	94·5	2-furyl ethyl ketone

Nominal mass	Composition	Mass measurement error
110	$C_7H_{10}O$	0·7 mMU
95	C_6H_7O	0·3 mMU

[a] On the basis of the mass measurement data and the strong indication of a furanoid structure from the search output, the compound was manually identified as ethyl methyl furan which was not in the library which was used (ref. 49).

The STIRS system (23) permits subsequent interaction by the operator to introduce further data, such as a suspected mol. wt, to see whether the consistency of the results from the different match factors is improved. The full STIRS search is, in fact, the second step in the identification procedure, having been preceded by a simpler matching procedure, so that only unknowns not identified in the first step are passed on to the second. A somewhat similar suggestion of eliminating known compounds by a matching procedure and using the much more costly computer interpretation procedure for the remaining unknowns had been made earlier by Biemann *et al.* (13a). Grönneberg *et al.* (19a) also suggested that their mini-computer based restricted file search could be followed by a large scale interactive file search for the remaining unknowns.

Finally, although some of the advantages and disadvantages of a large number of different search systems have been described in this chapter, it should be remembered that in an environment different from that used by the original author, e.g. a different library, perhaps covering a different mol. wt range, or perhaps using slightly different pre-search conditions, a system that was satisfactory in the original application may become unacceptable. No one library search system will meet all requirements.

REFERENCES

1. Abramson F. P. (1975). *Analyt. Chem.* **47**, 45.
2. Heller S. R. (1972). *Analyt. Chem.* **44**, 1951.
3. Ridley R. G. (1972). *In* "Biochemical Applications of Mass Spectrometry" (Ed. Waller G. R.), p. 177. John Wiley and Sons, Chichester.
4. (a) Dayringer H. E. and McLafferty F. W. (1976). *Org. Mass Spectrom.* **11**, 543.
   (b) "A Guide to Collections of Mass Spectral Data" (1974). (Ed. Middleditch B.S.). American Society for Mass Spectrometry.
   (c) "Eight Peak Index of Mass Spectra" (1974). (2nd Edition). Mass Spectrometry Data Centre, AWRE, Aldermaston.
5. (a) Jellum E. (1976). Personal communication.)
   (b) Crawford L. R. and Morrison J. D. (1968). *Analyt. Chem.* **40**, 1464.
6. (a) Knock B. A., Smith I. C., Wright D. E., Ridley R. G. and Kelly W. (1970). *Analyt. Chem.* **42**, 1516.
   (b) McGuire J. M., Alford A. L. and Carter M. H. (1972). 20th Annual Conference on Mass Spectrometry and Allied Topics, Dallas. Paper R6, p. 366.
7. "Registry of Mass Spectral Data" (1974). (Eds. Stenhagen E., Abrahamsson S. and McLafferty F. W.). John Wiley and Sons, New York.
8. McLafferty F. W., Busch M. A., Kwok K-S., Meyer B. A., Pesyna G., Platt R. C., Sakai I., Serum J. W., Tatematsu A., Venkataraghavan R. and Werth R. G. (1974). *In* "Mass Spectrometry and NMR Spectroscopy in Pesticide Chemistry" (Eds Biros F. J. and Hague R.). Plenum Press, New York.
9. Lawless J. G. and Romiez M. P. (1974). *In* "Advances in Mass Spectrometry" (Ed. West A. R.), Vol. 6, p. 143. Applied Science Publishers, Barking.

10. Mathews R. J. and Morrison J. D. (1974). *Aust. J. Chem.* **27**, 2167.
11. (a) Law N. C., Aandahl V., Fales H. M. and Milne G. W. A. (1971). *Clin. Chim. Acta* **32**, 221.
    (b) Akatsuka Y., Goto S., Noshiro M. and Yamagishi R. (1971). *Mass Spectroscopy* (Japan) **19**, 235.
    (c) Pettersson B. and Ryhage R. (1966). *Ark. Kemi* **26**, 293.
    (d) Tantsyrev G. D. and Kozlov S. T. (1971). *Analyt. Chem.* (U.S.S.R.) **26**, 588.
    (e) Abrahamsson S., Haggstrom G. and Stengagen E. (1966). 14th Annual Conference on Mass Spectrometry and Allied Topics, Dallas. Paper 105, p. 522.
    (f) Farbman, S., Reed R. I., Robertson D. H. and Silva M. E. F. (1973). *Int. J. Mass Spectrom. Ion Phys.* **12**, 123.
12. Cornu A. and Massot R. (1975). "Compilation of Mass Spectral Data" (2nd Edition). Heyden, London.
13. (a) Hertz H. S., Hites R. A. and Biemann K. (1971). *Analyt. Chem.* **43**, 681.
    (b) Costello C. E., Hertz H. S., Sakai T. and Biemann K. (1974). *Clin. Chem.* **20**, 255.
14. Hites R. A. and Biemann K. (1968). *In* "Advances in Mass Spectrometry" (Ed. Kendrick E.), Vol. 4, p. 37. Institute of Petroleum, London.
15. (a) Grotch S. L. (1973). *Analyt. Chem.* **45**, 2.
    (b) DaGrangano V. L. and Hotz H. P. (1974). *In* "Advances in Mass Spectrometry" (Ed. West A. R.), Vol. 6, p. 445. Applied Science Publishers, Barking.
    (c) Finkle B. S., Taylor D. M. and Bonelli E. J. (1972). *J. Chromat. Sci.* **10**, 312.
16. (a) Robertson D. H. and Merritt C. Jr. (1974). 22nd Annual Conference on Mass Spectrometry and Allied Topics, Philadelphia. Paper U7, p. 447.
    (b) Merritt C. Jr., Robertson D. H., Cavagnaro J. F., Graham R. A. and Nichols T. L. (1974). *J. agr. Fd Chem. 1974* **22**, 750.
17. Bell N. W. (1972). 20th Annual Conference on Mass Spectrometry and Allied Topics, Dallas. Paper R5, p. 364.
18. (a) Markey S. P., Urban W. G., Keyser A. J. and Goodman S. I. (1974). *In* "Advances in Mass Spectrometry" (Ed. West A. R.), Vol. 6, p. 187. Applied Science Publishers, Barking.
    (b) Hardy J. and Jardine I. (1974). *In* "Advances in Mass Spectrometry" (Ed. West A. R.), Vol. 6, p. 1061. Applied Science Publishers, Barking.
    (c) Mass Spectral Search System (MSSS), Cyphernetics International Corporation.
19. (a) Grönneberg T. O., Gray N. A. B. and Eglinton G. (1975). *Analyt. Chem.* **47**, 415.
    (b) Baty J. D. and Wade A. P. (1974). *Anal. Biochem.* **57**, 27.
    (c) Sweeley C. C., Young N. D., Holland J. F. and Gates S. C. (1974). *J. Chromatog.* **99**, 507.
    (d) Chapman J. R. and Street F. J. (Unpublished results).
    (e) Pesyna G. M., Venkataraghavan R., Dayringer H. E. and McLafferty F. W. (1976). *Analyt. Chem.* **48**, 1362.
20. Wangen L. E., Woodward W. S. and Isenhour T. L. (1971). *Analyt. Chem.* **43**, 1605.
21. van Marlen G., Dijkstra A. (1976). *Analyt. Chem.* **48**, 595.
22. Grotch S. L. (1970). *Analyt. Chem.* **42**, 1214.
23. Kwok K-S., Venkataraghavan R. and McLafferty F. W. (1973). *J. Am. chem. Soc.* **95**, 4185.
24. (a) Naegli P. R. and Clerc J. T. (1974). *Analyt. Chem.* **46**, 739A.
    (b) Erni F. and Clerc J. T. (1972). *Helv. Chim. Acta* **55**, 489.

25. McLafferty F. W., Dayringer H. E., Venkataraghavan R. (Unpublished data.)
26. Dayringer H. E., McLafferty F. W. and Venkataraghavan R. (1976). *Org. Mass Spectrom.* **11**, 845.
27. Dayringer H. E., Pesyna G. M., Venkataraghavan R. and McLafferty F. W. (1976). *Org. Mass Spectrom.* **11**, 529.
28. Mass Spectral Search System, Users' Manual (May, 1974). U.S. Environmental Protection Agency, Office of Planning and Management, p. 32.
29. von Unruh G., Spiteller-Friedmann M. and Spiteller G. (1970). *Tetrahedron* **26**, 3039.
30. Robertson D. H. and Reed R. I. (1971). 19th Annual Conference on Mass Spectrometry and Allied Topics, Atlanta. Paper D5, p. 68.
31. Rabinovich A. S. and Sakharov V. M. (1972). *Analyt. Chem.* (U.S.S.R.) **27**, 585.
32. Grotch S. L. (1971). *Analyt. Chem.* **43**, 1362.
33. Grotch S. L. (1969). 17th Annual Conference on Mass Spectrometry and Allied Topics, Dallas. Paper 171, p. 459.
34. Markey S. P., Urban W. G. and Keyser A. J. (1973). 21st Annual Conference on Mass Spectrometry and Allied Topics, San Francisco. Paper D2, p. 71.
35. Grotch S. L. (1975). *Analyt. Chem.* **47**, 1285.
36. Heller S. R., Koniver D. A., Fales H. M. and Milne G. W. A. (1974). *Analyt. Chem.* **46**, 947.
37. Hoyland J. R. and Neher M. B. (1974). E.P.A. Report-660/2-74-048, Office of Research and Development, U.S. Environmental Protection Agency, Washington.
38. Heller S. R., Fales H. M. and Milne G. W. A. (1973). *Org. Mass Spectrom.* **7**, 107.
39. Dayringer H. E. and McLafferty F. W. (1977). *Org. Mass Spectrom.* **12**, 53.
40. McLafferty F., Hertel R. H. and Villwock R. D. (1974). *Org. Mass Spectrom.* **9**, 690.
41. Pesyna G. M., McLafferty, F. W., Venkataraghavan R. and Dayringer H. E. (1975). *Analyt. Chem.* **7**, 1161.
42. "Olfax", Universal Monitor Corporation, Pasadena, California.
43. "Massmatch" system. Mass Spectrometry Data Centre, Aldermaston.
44. (a) Nau H. and Biemann K. (1974). *Analyt. Chem.* **46**, 426.
    (b) Nau H. and Biemann K. (1973). *Analyt. Letts* **6**, 1071.
45. Dromey R. G. (1976). *Analyt. Chem.* **48**, 1464.
46. Jellum E., Helland P., Eldjarn L., Markwardt U. and Marhöfer J. (1975). *J. Chromatog.* **112**, 573.
47. Jellum E. (1976). British Mass Spectroscopy Group Meeting, Liverpool.
48. Jellum E., Stokke O. and Eldjarn L. (1973). *Analyt. Chem.* **45**, 1099.
49. Technical publication T.P. 39 (1975). Kratos A.E.I. Scientific Instruments Ltd., Manchester.

6

Pattern Recognition

6.1 INTRODUCTION

The previous chapter discussed the identification of a compound from its mass spectrum by library searching. The assumption was made that another spectrum of the unknown was present in the libary and that the search would retrieve the library spectrum of the unknown as the most similar one. It was also pointed out that, if a spectrum of the unknown was not in the library, spectra of compounds with similar structures might often be retrieved, and, indeed some "interpretative" searches, e.g. the Self-training Interpretative and Retrieval System (STIRS), were designed to do exactly this. In this and the following chapter, we take a wider look at the problem of obtaining structural information by computer-based techniques from the mass spectrum of a compound for which a reference spectrum is not available. In particular, in this chapter, we look at pattern recognition techniques, addressed to the problem of providing classifications of unknowns based on mass spectral data.

A statement of the general problem addressed by pattern recognition techniques has been given by Kowalski and Bender (1). The statement asks: Can an obscure property of a collection of objects (in this case compounds) be detected and/or predicted using indirect measurements (in this case mass spectra), made on the objects, that are known to be related to the property via some unknown relationship? As an example we may wonder whether the presence of various functional groups may be detected from mass spectral peak intensity data.

A primary division of pattern recognition methods is into parametric and non-parametric. Parametric methods assume that the probability distributions of the data are either known or can be estimated. This has not usually been the case with mass spectral data and relatively few applications of parametric methods have been made in this field. However, their practical utility could be considerable. On the other hand, with non-parametric methods no assumption about underlying statistical distributions is made and the application of these methods constitutes the great majority of published work in pattern recognition in mass spectrometry.

If the measurements are related to the obscure property, then it is reasonable to assume that objects having this property in common, i.e. similar objects, will have similar measurements. A convenient representation that displays this fact is to consider the objects as points in a d-dimensional hyperspace, where d is the number of measurements made on each object. The values of the measurements become the co-ordinates of each point in d-space so that the more similar the objects, the nearer they cluster in d-space. If the property is known and if information regarding some objects with this property is given to the computer, the delineation of the cluster of objects with this property is called supervised learning. In this category, as major methods applied to mass spectrometry, we can place the calculation of distance from average spectra (Section 6.2), the use of learning machines (Section 6.3) and the K-nearest neighbour (KNN) method (Section 6.4). If the sought-for property is not exactly known, or if examples are non-existent, the application is one of unsupervised learning. Most unsupervised learning methods are actually cluster analysis methods (Section 6.6).

6.2 CLASSIFICATION BY MINIMUM DISTANCE FROM AN AVERAGE SPECTRUM

The simplest form of supervised learning is based on a minimum distance classifier. Classification by minimum distance from an average spectrum is appropriate in situations where each class can be represented by a single prototype point or pattern, P_i ($i = 1, 2, \ldots, R$), around which all other points (i.e. spectra) in that class tend to cluster. In this method, a point X, representing an unknown spectrum, is placed into that category, i, which is associated with the nearest of the average points, P_i ($i = 1, 2, \ldots, R$). Usually, the co-ordinates of each point, $X \equiv x_1, x_2, \ldots, x_d$ or $p_i \equiv p_{i_1}, p_{i_2}, \ldots, P_{i_d}$, are the intensity values at each of the d mass numbers in the mass spectrum. A common measure of distance is the Euclideean distance: $d_1(P_i, x) = [\sum_{n=1}^{d}(p_i - x_i)_n^2]^{\frac{1}{2}}$, or the square of this distance, $\sum_{n=1}^{d}(p_i - x_i)_n^2$, but other functions, e.g. $d_2(P_i, x) = \sum_{n=1}^{d}|p_i - x_i|_n$, which is more efficiently computed, can be used.

Naturally, the successful use of this technique depends upon the tightness of the cluster round the prototype pattern. The less tight it is, the less successful the method will be. Crawford and Morrison (2) investigated the properties of a number of such clusters and described them by means of an easily visualized geometrical method. In this method, the mass spectra are normalized so that the sum of the squares of the intensity values becomes one ($\Sigma I^2 = 1$) and thus the points representing the spectra lie on the surface of a hypersphere of unit radius. Because all the points in a cluster lie on this sphere, the centre of gravity of this group (see Fig. 6.1) will be inside the sphere. For a tightly knit cluster the centre of gravity is almost on the surface of the sphere; the greater the spread, the greater the distance (R) of this centre of gravity inside the sphere. These investigations showed, for example, that aromatic compounds form a more tightly defined group than do acids.

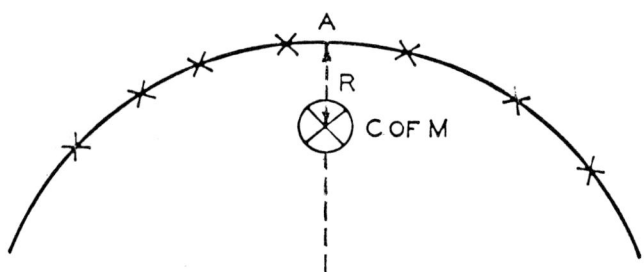

Fig. 6.1. The centre of gravity of compounds on the hypersphere used as a measure of the spread of the compounds over the surface. Distance R increases as the compounds become more scattered. (Reprinted with permission from ref. 2. Copyright by the American Chemical Society.)

The success of this method also depends on the degree of overlap between clusters it is wished to distinguish. This overlap was assessed by Crawford and Morrison by comparing first the distances between the projections of the centres of gravity of each group on the hypersphere, and second the distances between a given member of each group and these projected centres of gravity. These studies indicated that many clusters of simple aliphatic compounds were well defined and well separated when the common feature in a cluster was a functional group, although the degree of definition varied considerably between functional groups. Alternative groupings based on the number of rings, double bonds, oxygen atoms or nitrogen atoms were possible but, in general, although providing complementary information, were not as well defined.

As part of a wider study of computer-based identification procedures, in this case applied to the identification of alkylbenzenes from their mass spectra, Mathews (3) investigated a number of variants on classification by minimum distance from average spectra. These variants involved changes in the pre-processing of the data and the decision algorithm that was used. The data was pre-processed in one of three ways, (a) the spectra were standardized so that the total ion current was 1000 ($\sum I = 1000$), (b) after standardization to $\sum I = 1000$, the ion series spectrum (i.e. summation of intensities every 14 mass units) was constructed and used, (c) a new set of intensity values was formed from the Walsh nonlinear transformation of each spectrum

(60). The decision algorithm was either a straightforward minimum distance classifier or the same classifier with the decision surface shifted by an amount θ determined to give the best separation of the clusters. In these tests, the standardized spectra gave far better results (in terms of classification of unknowns) than did the ion series spectra and in both these categories considerable improvement was achieved by incorporating a value for θ into the decision algorithm (Table 6.1). The Walsh transform, previously introduced (60) to assist the classification of linearly inseparable data, and which was used in conjunction with the decision algorithm constant θ, gave results that were little better than the standardized spectra together with the constant θ. Thus, at least with this data, the previously reported improved performance of the non-linear transform method over that using standardized spectra (60) is due to the effect of the constant θ, rather than the non-linear transformation of the data.

TABLE 6.1
Structure assignment results[a]

Method	Number correct
Ion series	2
Ion series (after incorporation of θ)	16
Standardized spectra	9
Standardized spectra (after incorporation of θ)	34
Non-linear transform	36

[a] Results with 54 unknown alkylbenzenes using distance from an average spectrum method (see text for details). (From ref. 3.)

Work reported by Smith (4) on classification by minimum distance was however, more successful in using average ion series spectra. As can be seen from the following definition of a class used by Smith, the classes of data used were much more suited to ion series representation than the categories of alkylbenzenes considered by Mathews (3). Thus, the definition adopted was that for a general molecule: $R\text{-}(CH_2)_n H$ ($n = 0, 1, 2, \ldots$). "Compounds having R constant as n varies as indicated are considered to be in the same class if branches or extensions of the alkyl claim do not affect the basic fragmentation pattern by participating in rearrangements or other fragmentation pathways not normally available to lower homologues". Typical classes were "*n*-alkane", "branched alkane", "*n*-alkyl cyclohexane", "*n*-alkan-1-ol" and "2,6-isoprenoid methyl ester". The distance criterion used by Smith was a minimum value of $\sum_{n=1}^{d}|p_i - x_i|_n$. Refinements to these ion series classifications, particularly in cases where a number of possible classes had similar average ion series spectra were subsequently accomplished using specialized subroutines (see Fig. 6.2). An earlier study by Crawford and Morrison (2) on similar compounds led them also to favour the ion series approach.

```
                    Mismatch                  Class
    Phytane           5·7         Isoprenoid alkane
                      9·9         Highly branched alkane
                     18·9         Iso - alkane
                     20·0         n - Alkane
                     20·6         anteiso - Alkane
            Conclusion: acyclic alkane
            Molecular weight is 282
            Molecular formula is C20 H42
            Anomaly at mass 113 (M - 169)
            Anomaly at mass 127 (M - 155)
            Anomaly at mass 183 (M -  99)
            Anomaly at mass 197 (M -  85)
            Anomaly at mass 253 (M -  29)
            Anomaly at mass 267 (M -  15)
            Conclusion: probable isoprenoid alkane
```

Fig. 6.2. Summary of computer output for mass spectrum of phytane submitted as unknown. Mismatch values less than 50 indicate a good match. A branched alkane subroutine is then used to discriminate among the various classes of geochemically important acyclic alkanes (*n*, iso, anteiso) and isoprenoid by searching for anomalies or discontinuities in ion series. (From ref. 4.)

TABLE 6.2

Classification by distance from an average spectrum[a]

Chemical Structure	Predictive ability P						Reduced spectra
	Normal spectra			Difference spectra			
	lin	log	bin	lin	log	bin	
Double bond C=C	77·0	80·8	79·3	61·9	66·0	66·2	70·3
Hydroxysteroid	73·1	71·4	73·1	73·9	74·7	75·7	70·3

[a] Influence of data preprocessing (From ref. 5.)

Varmuza (5) studied the classification of 391 steroid spectra on the basis of the presence or absence of a number of structural features using the minimum distance from an average spectrum criterion. The effect of various forms of pre-processing of the data was also studied. Logarithmic or binary transformation of the intensity had little effect on the results. The use of difference spectra, i.e. losses from the molecular ion, gave poorer results except in the case of hydroxy-steroids. The use of ion series spectra with this data, practically always resulted in poorer predictive abilities (see Table 6.2).

Thus, we may note that an optimum data pre-processing routine for all classification problems does not exist. Again, no general theory that specifies one pre-processing routine rather than another in a particular case exists at present so that an empirical choice must be made to suit each problem. Further reference to pre-processing routines is made on p. 159.

6.2.1 Judging the Performance of a Classifier

Different parameters can be used to test the efficiency of a classifier. The predictive abilities, P_1 and P_2, are defined as the percentage of correctly classified spectra in each class and the reliabilities of answers, V_1 and V_2, as the percentage of "yes" and "no" answers that are correct (5). "Yes" is taken to indicate membership of category 1 which contains all compounds possessing a particular substructure; "no" indicates membership of category 2 which contains all other compounds. The predictive ability P, used by Isenhour et al. (6) is given by:

$$P = p_1 P_1 + (1 - p_1) P_2 = q_1 V_1 + (1 - q_1) V_2 \qquad (6.1)$$

where p_1 is the probability that a spectrum belongs to category 1 and q_1 is the probability of the answer "yes". An overall percentage prediction rate, such as P, rather than individual class figures such as V_1 and V_2 or P_1 and P_2 is not a good basis for comparison particularly if the rates for the two classes are dissimilar as may happen when one class is a minority class. It has also been pointed out that prediction rates can be subject to large statistical fluctuations (7). For example a 90% prediction rate from 100 test samples can indicate a true prediction rate of 82%–95% (even with 250 samples the range is 85%–93%) (8).

In a later paper, Rotter and Varmuza (9) introduced a single valued criterion, the information gain I, to judge the efficiency of a classifier. I is defined by the equation:

$$I = \sum_{i=1,2} \sum_{k=1,2} p(i, k) \ln \frac{p(i, k)}{p(i)\, p(k)} \qquad (6.2)$$

where $p(1, 1)$ is the probability of a spectrum belonging to category 1 being assigned to category 1, $p(1, 2)$ is the probability of a spectrum belonging to category 1 being assigned to category 2 and similarly for $p(2, 1)$ and $p(2, 2)$. As before $p(1)$ is the probability that a spectrum belongs to category 1 and $p(2)$ is the probability that a spectrum belongs to category 2.

An improvement in the practical applicability of minimum distance classifiers may be made by introducing the idea of rejections (cf. p.164) (5). Thus, a decision criterion, Δd, may be calculated as:

$$\Delta d = \frac{d_1 - d_2}{d_1 + d_2} \qquad (6.3)$$

where d_1 and d_2 are the Euclidean distances between the unknown and the average spectra of categories 1 and 2 respectively. If Δd lies between the thresholds of rejection, Δd_{min} and Δd_{max}, the spectrum is not classified; if

TABLE 6.3

Classification by distance from an average spectrum[a]

No.	Chemical structure	Percentage of spectra in category		V_1	V_2	P	Percentages of rejections for V_1 and V_2	
		1	2				$\geqslant 80\%$	$\geqslant 90\%$
1	Double bond C=C	48·6	51·4	74·0	80·3	77·0	15	45
2	Double bond C-4=C-5	26·3	73·7	63·9	94·4	83·4	15	35
3	Hydroxysteroid	70·8	29·2	89·4	52·6	73·1	40	43
4	Ketosteroid	73·9	26·1	89·5	62·9	81·6	13	25
5	Side chain at C-17	51·7	48·3	86·7	86·2	86·4	0	40
6	3-hydroxysteroid	38·6	61·4	74·1	81·9	79·0	10	70
7	3-ketosteroid	46·0	54·0	74·5	82·6	78·5	13	55
8	Oxygen function at C-3	90·0	10·0	97·3	34·4	82·9	21	21
9	Oxygen function at C-11	26·9	73·1	51·8	86·6	74·4	24	45
10	Oxygen function at C-17	59·6	40·4	87·6	75·3	82·1	25	45
11	5α-steroid	35·3	64·7	64·1	86·2	76·7	40	62
12	5β-steroid	23·5	76·5	43·8	87·4	72·1	35	75
13	OH in side chain at C-17	14·8	85·2	48·8	94·5	84·7	10	16
14	CO in side chain at C-17	26·6	73·4	78·5	93·0	89·0	4	11

[a] Influence of rejections on reliability. Results for "normal mass spectra" with linear intensities. (From ref. 5.)

$\Delta d \leqslant \Delta d_{min}$, the spectrum is assigned to category 1, if $\Delta d \geqslant \Delta d_{max}$, it is assigned to category 2. In practice, these rejection thresholds are altered until the reliability (V_1, V_2) reaches a desired value. For a number of categories, rejection rates as small as 10–20% increase the reliability significantly, although in other cases the method is of little value since rejection rates as high as 50–70% are required to produce a significant improvement (see Table 6.3).

6.2.2 Summary

In two papers (5, 10), Varmuza produces data to show that a basic deficiency of the minimum distance classifier is the weakly developed clustering in many of the systems studied. Thus, for some categories of steroid spectra, the distance between individual spectra and the corresponding average spectrum may be greater than the distance between the average spectra for the two categories (5). A further proviso is that satisfactory classifications cannot be expected by this or any other pattern recognition method unless a chemical structure is present in about 30–70% of the compounds used. Useful results can be achieved by distance measurement to average spectra although these results can be generally improved upon by use of learning machines or the KNN method. Distance measurement, however, has the advantage that no time-consuming training is necessary and the addition of new reference spectra is a relatively simple process.

6.3 CLASSIFICATION BY LEARNING MACHINES

Nilsson (11) defines a learning machine as "any device whose actions are influenced by past experiences". In this application we are dealing with linear learning machines that can be trained to recognize patterns (spectra) belonging to a particular class or cluster. Using the same geometrical representation in hyperspace (p. 151), any pattern (spectrum) of d elements, $X = (x_1, x_2, \ldots, x_d)$, where the x values are intensities at integer masses, can be represented as a $(d + 1)$ component vector (Y) in $(d + 1)$ dimensional space. The $(d + 1)$th component, whose value is always unity, is added so that a decision surface, passing through the origin, can be constructed to separate the two classes of spectra. The decision surface is a hyperplane with the equation $Y \cdot W = 0$, so that for those patterns on one side of the plane $Y \cdot W > 0$ and for those on the other $Y \cdot W < 0$. In this case, $W = (w_1, w_2, \ldots, w_{d+1})$, is called the weight vector. The discriminant function, $g(x) = Y \cdot W$, is implemented as a threshold logic unit (TLU), shown schematically in Fig. 6.3 and represented in eqn. 6.4:

$$g(x) = w_1 x_1 + w_2 x_2 + \ldots + w_d x_d + w_{d+1}. \tag{6.4}$$

Training the learning machine consists of finding a set of values for the weight vector, W, which allows it to correctly classify a given training set of spectra. The members of the training set are presented to the TLU one at a time. When a correct classification is made, no action is taken. Whenever the classification is incorrect, the weight values are adjusted so that the new weight vector will correctly classify the pattern. A correction method that has been found successful in much of the published work using mass spectra (32) is to define the corrected weight vector, W', so that $W'.Y = (W \pm cY).Y = -W.Y$ where the correction increment, c, is given by $c = \mp 2(W.Y)/Y.Y$. Thus, the misclassified point is moved across the hyperplane to a point an equal distance on the other side.

The correction process continues until all the patterns of the training set are correctly classified. It can be shown that if the categories are linearly separable, such a training procedure will converge towards a solution weight

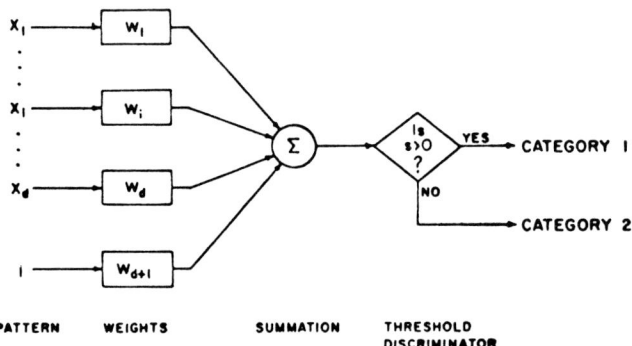

Fig. 6.3. Calculation and implementation of discriminant function. (Reprinted with permission from ref. 6. Copyright by American Chemical Society.)

vector (11). Usually, if the process does not produce a weight vector that will correctly classify all the training set after a pre-determined number of iterations through the training set, the training is terminated to conserve computer time. It has been customary to begin training with arbitrarily assigned weight values. However, it has been shown that a better procedure which speeds convergence is to use, where possible, a parametric discriminant function such as eqn. 6.12 to define an idealized initial vector (7).

The degree of success that can be achieved with learning machines in practice is strongly influenced by a number of factors. Three major factors may be mentioned, viz. (a) character of data, which may be altered by various pre-processing or feature extraction methods; (b) choice of categories to be separated and (c) choice of the training set from the spectra available. It should be noted that the influence of these factors is not limited to the use of learning machines so that many of the points raised here concerning these factors will be equally applicable to other pattern recognition procedures.

6.3.1 Character of Data—Pre-processing and Feature Extraction Methods

The character of the data may be altered by various pre-processing methods which can be useful for the following two reasons. First, some transformations can enhance information carried by the measurements. Second, pre-processing can reduce the dimensionality of the pattern space, either by discarding data unnecessary for the classification, or by combining data points. Reducing the dimensionality eases computational burdens and, if done correctly, will increase the chances of obtaining a meaningful classification through an increase in the ratio of the number of spectra available for training to the number of features (N/D ratio).

A reduction in the dynamic range of the peak intensities is commonly employed. This is carried out, whilst not rejecting any of the peaks in the original spectrum (usually covering an intensity range of 1%–100% or possibly 0·1%–100% base peak) by a transformation:

$$I'_n = f(I_n). \qquad (6.5)$$

where the most common functions are the logarithm or square root (32, 35). The choice of one of these methods can greatly improve the convergence speed in training. Other transformations which can enhance the information available in the spectra include the calculation of ion series spectra (12) or difference spectra (13, 5) or the combination of other highly correlating mass positions (14).

A form of alternate representation of mass spectra using moments rather than intensity values was introduced by Bender and co-workers as a means of dimensionality reduction (15). These authors have also introduced a method of constructing new variables that are optimal for the classification being undertaken, particularly for use with linearly inseparable data (16). Both of these methods will be discussed in the section on the KNN technique, to which they have been applied.

Studies of the Fourier (17) and Hadamard (18) transforms as pre-processing methods have demonstrated advantageous properties of the transformed data in each case. For example, the Fourier transformed data trained more rapidly and was less sensitive to deformations in the data than the original mass spectra. Both sets of transformed data allowed considerable reduction of dimensionality without an undue loss of classifier performance. However, neither of these transforms showed overwhelming advantages nor do they have a strong intuitive appeal as transformations of mass spectral data.

Where transformations produce new variables of widely differing magnitudes, some form of scaling is needed to counteract the inadvertently large weighting that measurements of large absolute magnitude would otherwise be given. Autoscaling (19) (see p. 173 for details) is a method that has been used in this situation and in the situation where data sets obtained from different techniques and expressed in different units are to be used together, e.g. the joint use of mass spectral and infra-red data in pattern recognition (20).

Weighting of measurements to increase their contribution can be used after scaling and, provided a reasonable basis for calculation of weights is available, can lead to improved results. In an application to gasoline analysis, Tunnicliff and Wadsworth (21) (p. 167) weight the data from training set compounds according to the percentage of that compound present in typical gasoline samples. Variance-based weighting factors that emphasize features that contribute most to class separability, such as the Fisher ratio, R in eqn 6.6, have been used by some authors (19, 22):

$$R = \frac{(m_1 - m_2)^2}{\sigma_1^2 + \sigma_2^2}. \qquad (6.6)$$

Processes for selectively removing data or features, not necessary for the classification being attempted, have been reported by Jurs and coworkers in a number of publications (23, 24, 25). The methods have been steadily refined. In the first method reported (23), a fixed number of m/e values (usually 15) with the lowest products of weight and intensity ($w_n x_n$ in eqn. 6.4), i.e. those making least contribution to the decision process, were dropped. About half the m/e values could be removed before the classification performance fell appreciably. A later method (24) relied on the fact that different weight vectors may be trained to dichotomize the same data set, for example by starting with all components of the weight vector initialized at $+1$ or all at -1. After training of these two vectors, m/e positions for which the signs of the components of the trained weight vectors differ are considered ambiguous and are discarded. Training is repeated with the reduced data set and the process is continued until no more ambiguous m/e positions are found. Some improvement in classification performance with the reduced data set was noted even though the final number of features was only one-quarter of the original. A more sophisticated version of this process has recently been applied to linearly separable data (25). In this case, a series of weight vectors is trained for the classification and the relative variation of these values for each weight is determined. Features having corresponding weights with the greatest variance are then removed until linear separability is just maintained. This method also ranks the relative importance of the features and can be used to estimate the intrinsic dimensionality of the data set, i.e. the minimum number of dimensions needed to effect linear separability.

A number of the features that remained after this feature extraction process, could reasonably well be correlated with the class description on the basis of the known fragmentation mechanisms of that class. For example, of the 14 m/e values strongly correlating with oxygen presence (24), the series m/e 31, 45, 59, 73 ($H(CH_2)_nO-$ $n = 1,...,4$) was prominent.

A more empirical method of data reduction has been used by Mathews (26). The problem was to train a vector to identify phosphonates (Formula I) from their mass spectra:

$$(RO)_2P(=X)Y. \quad (I)$$

R is H, CH_3 or C_2H_5; X is O or S; Y is any functional group. One training set represented a wide range of phosphonates, the other represented non-phosphonate compounds, chosen because of their structural similarity to the side chains of phosphonates. The average spectrum of each class was computed and only masses that were strong in either one average spectrum and weak in the other were retained. These were retained either as masses, mass ranges or summed masses—only eight features in total (Table 6.4). This more "chemical" approach was successful in this case because the fragmentation patterns of phosphonates are sufficiently well differentiated from those of most other organic compounds with ions at fairly uncommon m/e values (e.g. m/e 47 (PO^+), m/e 79 ($CH_3PO_2H^+$), m/e 81 ($PO_3H_2^+$)). The

TABLE 6.4

Eight feature spectrum classifier for phosphonates[a]

Pattern feature	m/e values in feature	W_i
1	47	1·77
2	50–59	−0·50
3	66–76	−0·23
4	79, 81	1·61
5	85, 86	−0·96
6	109, 125	0·24
7	132, 133, 137	0·60
8	136, 139, 140	−0·81

[a] $S > 0$ for category 1 (phosphonates) and $S \leq 0$ for category 2 (non-phosphonates), where $S = \Sigma_{i=1}^{8} X_i W_i$ and X_i is the intensity of the most intense peak of the m/e values listed for a particular pattern feature. This classifier correctly classified all of the phosphonates in the test set (34 compounds) and 63 of the 76 non-phosphonate compounds in the test set. (From ref. 26.)

"chemical" approach was also recognized at a very early stage by Raznikov and Talroze (27) in pioneering work on pattern recognition. These authors used pattern recognition procedures to develop a more refined relationship to distinguish paraffins and olefins using only the known characteristic masses, m/e 41, 43, 55 and 57. Further studies using a very restricted set of masses to distinguish monoolefins and cycloparaffins were also made by these authors.

An interactive feature selection method based on KNN analysis has been described by Pichler and Perone (28) and used by Burgard et al. (29) to select features suitable for the classification of fragments from the mass spectra of oligonucleotides. A variance based method of feature selection has been used by Vink and coworkers (30).

6.3.2 Choice of Categories

Using mass spectra as the data base, most studies have attempted to classify compounds on the basis of structural features that are more obviously related to the spectra. Care must be exercised in the choice of these categories, since no learning machine implementation can overcome a poor choice of categories, i.e. categories where no suitable relationship to the mass spectral data is available.

An excellent example of the effect of choice of category is given in the work of Mathews describing two of the learning machine schemes used for the identification of alkylbenzenes (3). In one scheme, the "presence of structure" (PSLM) method; weight vectors relating to the presence of each possible side chain structure were trained for each molecular weight. Each unknown with the same mol. wt was then tested with the appropriate vectors and the unknown constructed by adding the side chains with positive response to a benzene nucleus to give a structure of the correct mol. wt. The number of training iterations for the identification of compounds with mol. wt 134 is given in Table 6.5. The other scheme is based on a chemist's flowchart (31) which asks the questions that, from a knowledge of the fragmentation processes of alkylbenzenes, are most readily answered using the mass spectrum. A flow chart for this method is shown in Fig. 6.4 and the number of training iterations required by this method listed in Table 6.5.

As can be seen from Table 6.5, all the chemist's flowchart questions are linearly separable, whereas three of the eight PSLM questions are not. By using the learning machine to produce weight vector values for the questions in the flowchart, its predictive ability was increased well above that of the original scheme where such values were manually adjusted to best fit the training data. As Mathews points out (3), this "chemist-learning machine" interaction achieved results that are considerably better than the results of

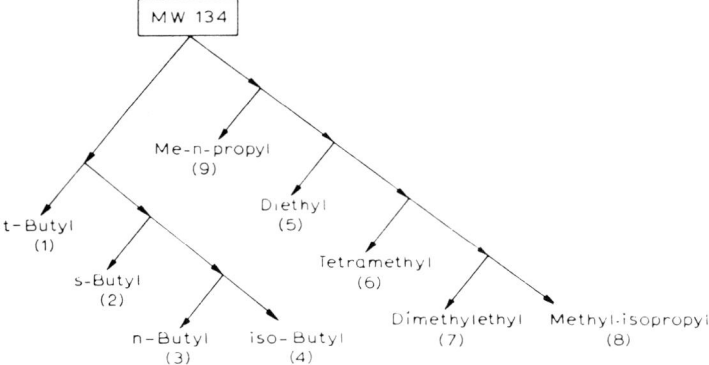

Fig. 6.4. "Chemist's flowchart" for alkylbenzenes with a mol. wt of 134. (From ref. 3.)

TABLE 6.5

Number of training iterations required for the identification of alkylbenzenes of mol. wt 134

Presence of structure	No. of iterations	Flowchart question (Fig. 6.4)	No. of iterations
n-Bu	5	(1–4) v. (5–9)	5
iso-Bu	5	(9) v. (5–8)	1
sec-Bu	12	(5) v. (6–8)	1
t-Bu	2	(6) v. (7–8)	1
n-Prop	10	(7) v. (8)	81
iso-Prop	>100	(1) v. (2–4)	1
Et	>100	(2) v. (3–4)	1
Me	>100	(3) v. (4)	1

[a] "Presence of structure" method and chemist's flowchart method. (From ref. 3.)

either the chemist or the learning machine alone. It may well be that a learning machine will be able to achieve satisfactory results with other types of "linearly inseparable" data if such a chemist-learning machine interaction can be used to optimize the relevance of the questions being asked.

A further point with regard to choice of categories is that a reliable classification of unknown spectra cannot be expected if the appropriate category is represented by less than about 30% of the spectra used in training, even though the training procedure may converge to give perfect recognition of the training set (5, 7).

6.3.3 Choice of Training Set

In general, it is desirable to use the largest, most representative training set available. As has been pointed out in refs 7 and 44, apparently perfect separation on training can almost always be obtained if the ratio N/D, i.e. the total number of spectra in both classes of the training set, divided by the number of features in the spectra is equal to or less than two. However, under these circumstances, the separation is quite possibly meaningless for the classes under consideration. The ratio N/D should be greater than three for meaningful results (7, 15). Thus, the use of large training sets and feature reduction techniques is to be recommended. Again, the proviso that neither of the training sets for the individual classes should be so small as to be unrepresentative should also be noted.

6.3.4 Linear Separability

When faced with linearly inseparable data, i.e. when the training process does not converge, it has been usual to stop the training process after either a specified time or a certain number of iterations (33). However, with learning machines, the final position taken up by the hyperplane and thus the performance of the resulting classifier is heavily dependent on the last error correction made before training was stopped, and can be drastically altered, for example, by changing the order of the data in the training set. Bender and Kowalski (34) have found the linear learning machine to be highly unreliable when using inseparable data.

A number of approaches already discussed may help to achieve separable data, e.g. pre-processing and feature extraction, careful choice of categories and the use of a rejection zone or deadzone (35, 36). For a learning machine, the rejection zone is, in effect, a separating plane that has a finite width. Thus, in prediction, if $Y.W > \Delta$, the pattern belongs to one category, if $Y.W < \Delta$, then it belongs to the other category, and if $-\Delta < Y.W < \Delta$, the pattern is not classified (Fig. 6.5). In training, feedback is applied to correct both an incorrect classification and a failure to make a classification.

Unfortunately, as previously mentioned, although this method usually improves the reliability of the classifications, the actual number of classifications made is reduced by the existence of the rejection zone. In this respect, it is interesting to note that in a study where some less important m/e values were deleted and replaced by cross terms, i.e. terms that refer to relationships between masses, then for almost every data set and classification considered, the number of rejections was reduced by the inclusion of these cross terms (14). Obviously, this form of data pre-processing can help to increase the degree of linear separability of data.

Another approach to achieving linear separability was used by Mathews (26). In training a vector to separate phosphonates, from a general data set representing non-phosphonates (p. 161), those members of the non-phosphonate category causing the inseparability were detected by inspection

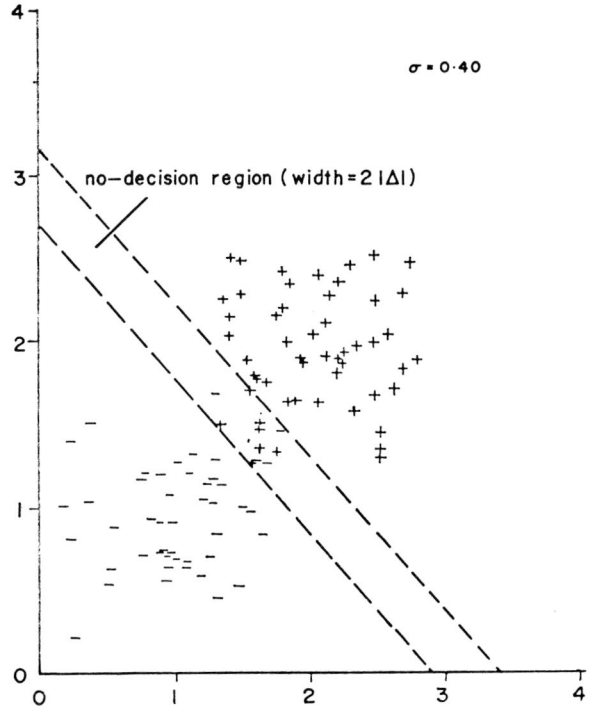

Fig. 6.5. No-decision region for inseparable data. (Reprinted with permission from ref. 36. Copyright from American Chemical Society.)

of the error corrections and removed from the training set. In this case, transfer of 5 of the 68 non-phosphonate compounds in the training set to the prediction set was sufficient to allow linear separability. This type of procedure may not always be so readily applicable, but it is obviously of value to try to locate the particular patterns causing linear inseparability when it occurs.

6.3.5 Multicategory Prediction

While a number of chemically significant questions, such as the presence or absence of a certain functional group, can be answered by a binary decision as implemented by a simple TLU, multicategory decisions, such as the number of certain groups or atoms per molecule, must also sometimes be made. One means of achieving multicategory classifications is through the use of an array of binary classifiers. One form of array is the branching tree (e.g. Fig. 6.4). The branching tree method has been used as the basis of a scheme for molecular formula determination from low resolution mass spectra (6). Another form of array is the parallel arrangement (37). Both the branching and parallel systems and a binary coding system in which each decision (0 or 1) supplies a digit in a binary version of the category number have been compared in their application to the determination of the carbon number of low mol. wt compounds (38).

An alternative technique for multicategory decisions uses binary decisions to separate each class from each of the other classes (39). A large number of weight vectors is required; $n(n-1)/2$ for n classes. The category is determined by a majority vote. For example, in Fig. 6.6 a pattern in class A is classified as "A" by separator I, "A" by separator II and "B" by separator III. Hence the classification is "A" by majority vote.

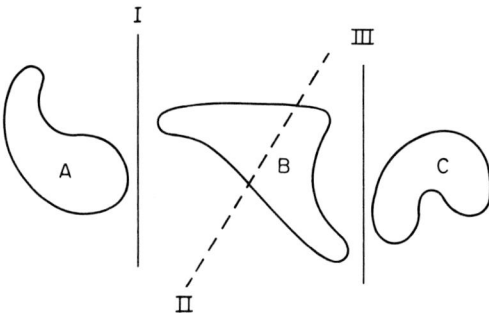

Fig. 6.6. Classes (A, B, C) which are not linearly separable. (Reprinted with permission from ref. 39. Copyright by American Chemical Society.)

Another aspect of the multicategory problem is the prediction of a continuous valued property, e.g. mol. wt (23), rather than separation of patterns into discrete classes. An approach to this problem using learning machines is to compute a weight vector, W, such that the dot product $Y \cdot W$ takes up a value, S, for each compound which is as near as possible to the magnitude of the property being determined. A least squares procedure, equivalent to the usual feedback training, is used to calculate the individual weights, w_j, using reference compound data, so that the expression $(S - S^*)^2$, where $S = \sum w_j y_j$ and S^* is the correct value for that compound, is a minimum (23). This least squares procedure is also applicable to more conventional multi-category problems, e.g. the determination of oxygen number.

Since its introduction, the least squares procedure has been tested as an important and practical method for the analysis of hydrocarbon types and the average molecular structure of gasoline (21). In this case, the dot products, S, are calculated to give a quantitative measure of the presence of types such as C_8 benzenes or structures such as side chain double bonds in the gasoline. In training, the dot product is set equal to 100·0 for all reference compounds belonging to the type being determined and 0·0 for all compounds not of that type. In the gasoline analysis, the dot product expresses the presence of a particular type as a volume percentage figure, whereas structures are calculated as the number of structural groups per average molecule. Table 6.6 shows the results for type analysis of four gasoline blends. The results in the column headed "Exptl data" are based on the use of experimentally measured mass spectra from the gasoline. The results in the column headed "Calcd data" are based on the use of mass spectrometric data calculated from reference spectra in the training set and the known composition of the sample. Weighting factors, reflecting the percentage of each training compound present in a typical gasoline, were used to obtain the data in Table 6.6 and were shown to effect a particular improvement in the agreement between calculated and theoretical data. This was believed to indicate that the principal errors remaining were then due to discrepancies between the training set spectra and those obtained using the authors' instrument. In any case, the data in Table 6.6 shows the excellent results that may be obtained by this method in practice.

Finally, it should be pointed out that the KNN method is a multi-category method without modification.

6.3.6 Digital Learning Nets

A somewhat different approach using adaptive digital learning nets has been adopted by Stonham et al. (40). The nets used consist of 128 Stored Logic Adaptive Microcircuit (SLAM) elements (Fig. 6.7). These SLAM elements may be implemented in hardware or, as in this case, simulated in software on a digital computer. Each element has address inputs which are connected at random to 4 bits of input data from a binary encoded spectrum. The same connections are used in both the training and recognition phases.

Information can be stored in each element to enable the device to give a pre-determined output for a given combination of inputs. This response is

TABLE 6.6
Results from type analysis of gasoline using weighted mass spectrometric data[a]

Hydrocarbon type	Theory	131 Exptl data	131 Calcd data	Theory	132 Exptl data	132 Calcd data	Theory	133 Exptl data	133 Calcd data	Theory	134 Exptl data	134 calcd data
0 Rings + double bonds	43·6	42·0	43·3	35·2	34·0	34·7	45·5	44·7	45·2	26·6	27·1	26·1
1	11·2	10·4	10·9	30·2	30·2	30·7	19·0	18·0	18·8	47·4	45·6	48·3
2	0·0	−0·2	−0·1	0·0	−0·2	−0·1	0·0	0·0	−0·1	0·0	−0·1	−0·1
4	42·9	44·5	42·9	32·7	33·6	32·6	34·0	35·3	34·0	24·0	25·2	23·8
5	1·5	1·7	1·5	1·4	1·5	1·3	1·0	1·1	1·0	1·4	1·7	1·5
7	0·7	0·7	0·7	0·5	0·6	0·6	0·4	0·5	0·4	0·6	0·7	0·6
Total	99·9	99·1	99·2	100·0	99·7	99·8	99·9	99·6	99·3	100·0	100·2	100·2
Total C_6	23·3	21·1	23·3	24·8	22·2	25·5	29·7	26·7	30·0	34·3	32·8	35·2
C_7	21·3	25·7	23·3	24·2	28·9	26·2	21·8	25·5	23·4	24·2	25·6	24·8
C_8	27·8	23·3	27·0	29·4	27·1	29·1	25·2	22·4	24·6	24·9	24·1	24·3
C_9	17·6	19·8	17·7	13·6	13·2	13·1	15·2	15·2	14·9	10·1	8·1	9·4
C_{10}	8·9	9·3	9·1	7·0	7·4	7·1	7·2	8·1	7·0	5·6	7·1	5·6
C_{11}	1·1	0·9	0·8	0·9	0·8	0·4	0·9	0·8	0·5	0·9	0·3	0·5
C_{12}	0·0	−2·2	−1·8	0·0	−2·5	−2·0	0·0	−2·2	−1·6	0·0	−2·9	−1·7
Total	100·0	97·9	100·1	99·9	97·1	99·4	100·0	96·5	98·8	100·0	95·1	98·1
Benzene	1·9	1·9	1·9	1·4	1·5	1·4	1·5	1·6	1·5	1·0	1·1	1·1
Toluene	8·2	8·4	8·5	6·3	6·4	6·5	6·5	6·7	6·7	4·6	4·8	4·8
C_6 Benzenes	14·3	14·3	14·2	10·9	10·7	10·7	11·3	11·4	11·2	8·0	8·0	7·8
C_9	12·6	13·3	12·1	9·6	9·7	9·2	10·0	9·8	9·7	7·0	7·3	6·8
C_{10}	6·0	6·3	6·2	4·5	5·0	4·8	4·7	5·2	4·7	3·3	3·9	3·3
C_{11}	0·0	0·3	0·1	0·0	0·3	0·0	0·0	0·4	0·0	0·0	0·1	0·0
C_{12}	0·0	−0·1	−0·2	0·0	−0·1	−0·3	0·0	−0·1	−0·2	0·0	−0·3	−0·2
Total	43·0	44·4	42·8	32·7	33·5	32·3	34·0	35·0	33·6	23·9	24·9	23·6

	1	2	3	4	5	6	7	8	9	10	11	12
Total olefins	0·0	−1·7	−1·4	0·0	−1·6	−0·9	8·0	5·6	7·6	16·4	14·2	17·3
Monoolefins	0·0	−1·7	−1·9	0·0	−1·6	−1·4	8·0	5·6	7·3	16·4	14·5	17·4
Monoolefins C=C in chain	0·0	−2·7	−2·2	0·0	−2·5	−1·6	8·0	5·1	7·5	16·4	14·2	18·1
Monoolefins C=C in ring	0·0	0·1	−0·5	0·0	0·0	−0·6	0·0	0·0	−0·6	0·0	0·2	−0·6
5-Carbon-ring sats.	6·9	6·3	7·1	15·4	20·2	21·1	6·9	7·9	6·9	15·4	20·0	21·7
5-Carbon-ring total	7·4	5·5	6·9	15·7	18·2	19·6	7·2	6·6	6·6	15·8	18·2	20·2
6-Carbon-ring sats.	4·3	3·6	3·1	14·9	14·8	13·7	4·1	3·6	3·0	15·6	16·3	14·5
6-Carbon-ring non-aromatic	4·3	3·7	2·9	14·9	15·0	13·6	4·1	3·7	2·8	15·6	16·7	14·5
6-Carbon-ring total	49·4	52·1	49·5	49·4	50·4	47·7	39·6	41·4	39·2	41·6	43·1	39·4
Total 1-ring sats	11·2	9·9	10·2	30·2	31·6	31·4	11·0	11·2	10·1	31·0	33·1	32·8

[a] (Reprinted with permission from ref. 21. Copyright from American Chemical Society.)

determined during the training phase when the desired response (a "1") to all 4 bit patterns seen by that element is stored in the appropriate location. One such learning net is required for each class; classification of an unknown is made according to which net gives the maximum response summed over all its elements. Although most work has been carried out with random connection mapping, the system can be improved by deliberate modifications to the connections to reduce the amount of redundant processing in the networks (41). So far, results have shown high recognition rates in the classification of spectra by functional groups. The technique is not limited to the classification of linearly separable data and can offer savings in time and storage requirements over other learning machine approaches.

6.3.7 Summary

Learning machines have been applied to a range of classification problems using mass spectrometric data (3, 5, 6, 12, 21, 23, 26, 27, 37, 42) and as part of more complex schemes of structural analysis (13). Structural analysis using infra-red and other data in addition to mass spectrometric data with learning machines has also been attempted (20). Papers by Jurs *et al.* (43) have used learning machine classification procedures in reverse to enable mass spectra to be predicted for selected compounds on the basis of their structures.

The major drawback of learning machine methods lies in the ease with which they can be misused so as to produce apparently perfect but meaningless separations of data which is then used for predictions (44). In addition, the learning machine can be highly unreliable when used with linearly inseparable data (34). Nevertheless, especially with carefully chosen categories and features, the technique can yield valuable results. The chemist-learning machine interactive approach (p. 162) (3, 26) is especially to be commended in this respect. Another promising application of these methods is the quantitative analysis of very complex mixtures (p. 167) (21). The computer time and storage needed in learning machine training are in general, high, and training has to be repeated when the training set is enlarged. On the other hand, the subsequent use of the learning machine data for classification is computationally very undemanding.

6.4 CLASSIFICATION BY *K*-NEAREST NEIGHBOUR METHOD

In the K-nearest neighbour (KNN) method, an unknown pattern is classified according to the class of the nearest K patterns in the data set. When $K > 1$, the classification is that of the majority of the KNNs. The coordinates of the pattern X, representing the unknown spectrum $(X = x_1, x_2, \ldots, x_d)$ are usually the intensity values at each of the d mass numbers in the mass spectrum and the usual measure of distance is the Euclidean distance. The KNN method is a multi-category procedure without modification (cf. p. 167). Any number of categories can be represented by patterns in the data set and an unknown will still be classified according to a majority vote. Again, the KNN method provides much better classification results with linearly inseparable data than does the linear learning machine (34). This is because the KNN procedure always generates a unique classification unlike the linear learning machine which, in the case of linearly inseparable data, produces a classification that is heavily dependent on the last error correction, made before training was stopped (cf. p. 164).

Extensive assessments of learning machines and the KNN procedure using the same data set and categories have been made by Varmuza et al. (45, 46). The classifications studied were two class problems where each classifier was trained to determine the presence or absence of a particular chemical substructure. Little difference could be seen in the results for 31 such substructures based both on assessment of predictive ability (P) and on reliabilities (V_1 and V_2) when using learning machines or the KNN method with $K = 1$. However, it should be noted that all but one of the 31 categories in this example were linearly separable. Further investigations on a smaller number of these categories with the KNN method showed that increasing the number of neighbours ($K = 6$) and use of a majority vote afforded a marked improvement in the results, with highly reliable identifications of the substructures sought (10). For example, 99% of the predictions of the presence of oxygen in a molecule were correct as were 86% of the predictions of the presence of a keto-group (see Table 6.7). The parallel between this method and a library search using a distance metric where, although the unknown is not in the library, the fact that a number of the best matches contain a common substructure suggests that the unknown also contains this substructure should be noted.

In other references, the use of the simple one nearest neighbour method, i.e. $K = 1$, at least in initial trials of the method, has been generally recommended (34, 47). The studies of Justice and Isenhour (47) also concluded that the nearest neighbour method was the pattern recognition method giving the highest predictive ability. Other methods tested by these authors included the learning machine and distance from average spectrum approaches.

One drawback of the KNN method is the large amount of computing time required for classification if every library spectrum is brought down from a backing store and used in distance calculations. However, it should be possible to reduce this time by comparing an unknown with standard prototypes and then retrieving for individual comparison only those refer-

TABLE 6.7
Comparison of learning machine and KNN classifications[a]

Class	Overall % of class	6NN		1NN		Learning machine		% of class in prediction set
		V_1	V_2	V_1	V_2	V_1	V_2	
Alkane	8	75	99	76	98	65	99	7
Aliphatic alcohol	12	88	93	58	94	54	97	11
Benzene ring	23	96	99	88	97	91	97	22
Pyridine ring	7	77	97	53	97	56	98	6
Carbonyl group	28	86	79	67	85	64	84	30
Phenol	4	82	98	71	99	80	99	4
Methyl group	78	85	73	86	62	89	55	78
Ethyl group	40	67	82	61	75	64	70	43
Oxygen present	53	99	77	86	78	84	83	55
Nitrogen present	25	94	90	80	92	78	95	24

[a] All 500 spectra were used for the KNN assessment. A prediction set of 250 was chosen from these to assess the learning machine classification. Intensity values were logarithmic, normalized to the base peak. V_1 = reliability of answer "yes". V_2 = reliability of answer "no". (Data from refs 10, 45 and 46.)

ences close to matching prototypes (48). For the same reason, considerable incentive to abbreviate spectra as much as possible when using the KNN method exists. In addition to abbreviation methods described previously (Section 6.3.1), two other methods used by Kowalski and co-workers in their investigations of the KNN method can be mentioned. One of these is to use moments rather than intensities for each m/e value (15, 16). 10 X_1–X_{10}, were defined as variables for each spectrum as follows:

$$X_1 = A\Sigma W_i v_i \qquad X_k = A\Sigma W_i(v_i - X_1)^k \qquad k = 2, \ldots, 5 \qquad (6.7)$$

$$X_6 = B\Sigma W_i v_i \qquad X_{5+k} = B\Sigma v_i(W_i - X_6)^k \qquad k = 2, \ldots, 5 \qquad (6.8)$$

$$A = 1/\Sigma W_i \qquad B = 1/\Sigma v_i.$$

where v_i and W_i are the m/e value and the square root of the intensity respectively for each peak of non-zero intensity. This alternative representation with only 10 variables resulted in no more than a 5% loss of classification performance. A further transformation discussed by Bender and Kowalski (16) that defines optimum variables for a particular classification was applied to moment data and resulted in an increased classification performance with a reduction to only three variables. Bender and Kowalski tested these data transformations using three classes of randomly selected hydrocarbon spectra; the spectra within each class were of compounds containing six, seven and eight carbon atoms respectively. For each class, 40 spectra were used in the training set and 10 were used for the evaluation set. Results obtained with these spectra are shown in Table 6.8.

TABLE 6.8

Classification performance of KNN method[a]

	10 dimensions	3 dimensions
1NN		
Training set	102/120	113/120
Evaluation set	23/30	28/30
3NN		
Training set	96/120	110/120
Evaluation set	18/30	29/30

[a] $k = 1$ and $k = 3$ with 10- and 3-dimensional data from hydrocarbon spectra. (Reprinted with permission from ref. 16. Copyright by the American Chemical Society.)

In transformations such as the calculation of moments or optimum variables where the resulting variables have widely differing magnitudes or are measured in different units, a weighting of the measurement with the largest absolute magnitude will be inadvertently applied. To give the measurements an equal weight, some form of scaling such as autoscaling is applied (19). In this procedure, the measurements are scaled so that they each have a mean of zero and unit standard deviation. The kth coordinate of the ith pattern then becomes:

$$Y'_{ik} = (Y_{ik} - \overline{Y}_k)/\sigma_k$$

where

$$\overline{Y}_k = 1/N \sum_{i=1}^{N} Y_{ik}$$

and

$$\sigma_k^2 = \sum_{i=1}^{N} (Y_{ik} - \overline{Y}_k)^2.$$

Y_{ik} is the original coordinate and N is the total number of patterns to be autoscaled. Pre-processing by a technique such as autoscaling is also necessary where data collections from different techniques and expressed in different units, are to be used together, e.g. the joint use of mass spectral and infrared data in pattern recognition.

A recent application of the KNN method is contained in a strategy for the sequencing of deoxy-oligonucleotides by pattern recognition analysis of the mass spectra of the underivatized molecules (29). In the mass spectrometer, such molecules are reproducibly cleaved to give fragments which can be used to reveal base sequence information on the original oligonucleotide. An interactive feature selection method based on KNN analysis (28) was used to select features (ratios of specific ions or combinations of these) which were best able to distinguish the various sequences. Thus, three ion ratios in combination could be used to predict, with 100% accuracy, the presence of one of any of the 10 possible types of linkage between two nucleotides.

6.4.1 Summary

Probably more is known about how the KNN method works and how well it can perform than any other non-parametric classification procedure (34). As a result of this, it has been suggested as a standard by which other classification procedures may be judged (49). Although it can make considerable demands on computer time and storage, especially if full spectra are used, the facts that no training is required and that new spectra can be added to the library without costly updating, are in its favour. In addition, the generally superior results obtained and its direct applicability to multicategory situations presently recommend KNN as a more satisfactory classification method than the use of learning machines in most cases.

6.5 PARAMETRIC CLASSIFICATION METHODS

Parametric methods use the known or training data set to establish estimates of probability distribution parameters that characterize the data. These estimated parameters are then used to specify the decision function required. An example of this approach to pattern recognition is the work of Franzen (50). Considering a single mass, m, the frequency distribution of ion intensity at this mass can be plotted for each of the two classes to be distinguished. In general, two different distributions result (see Fig. 6.7).

For an unknown spectrum, with an intensity $I_{u,m}$ at mass m, a probability ratio P_1/P_0 can be read off to give an indication of the likelihood of the unknown belonging to one class rather than the other. Thus, a decision function can be defined:

$$S_m = \log r_m = \frac{\log P_{1,m}(I_{u,m})}{\log P_{0,m}(I_{u,m})} \tag{6.10}$$

where $S_m \geq 0$ for class 1 and $S_m < 0$ for class 0. Considering all masses, the discriminant function becomes:

$$S = \sum_m S_m \tag{6.11}$$

where $S \geq 0$ for class 1 and $S < 0$ for class 0.

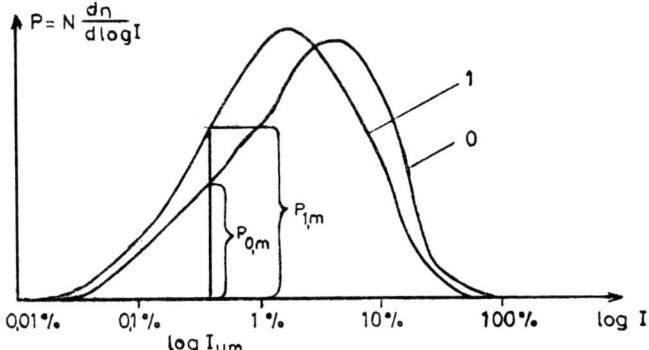

Fig. 6.7. Example of normalized frequency distributions of the intensities of all learning set spectra at mass m. The two curves, designated with 1 and 0, represent the spectra with presence or absence of the structural detail considered. (From ref. 50b.)

The use of logarithms allows this simple addition of probabilities. Franzen and Hillig (50b) investigated two methods of deriving discriminant function data from training data distributions. One was to assume that the distributions were Gaussian and describe them by their mean and variance. The other was to store probability ratios for a limited number of intensity ranges, chosen to have roughly equal populations, for each mass. In practice, the latter method, using only four to six intensity ranges, yielded the better discriminant functions. In either case, the assumption is made that the intensities at different masses are independent. This is not so. This assumption could be avoided by the use of a transform, such as that applied in Factor analysis (p. 182), to create new, independent variables. By this means, benefit would also be gained from a reduction in the number of variables and hence the number of values of S_m that must be stored. However, at present it is not certain what effect the assumption of independence has on the results.

The calculated value of S not only classifies the unknown, but also offers a measure of classification reliability. If the value of S is too small to give an acceptable reliability, a second decision may be used to improve this figure (50a). The S_m values for this second decision are calculated only from those spectra that give correspondingly small values in the first decision. Some results from the two methods are compared in Table 6.9.

TABLE 6.9

Success rates with parametric method of Franzen[a]

Category	1 step method	2 step method
Aromatic	95·6%	98·9%
Oxygen present	87·7%	95·0%
$n(H) > 2 \cdot n(C)$	86·8%	92·6%
Nonaromatic ring	84·5%	90·2%

[a] (From Ref. 50a.)

Another application of parametric methods to pattern recognition appears in an excellent paper by Vink and coworkers (30). Following an earlier study on the recognition of stereoisomeric monosaccharides (61), this work applies parametric methods to the identification of some stereoisomeric N-acetylhexosamines from the mass spectra of their trimethylsilyl derivatives. Differences in the spectra of these stereoisomers exist but are so small as to be almost unusable by normal methods (see Fig. 6.8). Some improvement is possible if intensity ratios are used, but even so, the use of pattern recognition methods is necessary.

Suitable ratios were chosen by the following means: (a) Mass values below m/e 147 were not taken into account to avoid interference from background present in the low mass range. (b) Only peaks which were obviously not isotope peaks and occurred in each spectrum with an intensity of at least 0·5% with respect to the base peak (m/e 173) were used. (c) Only peak intensity ratios of which the corresponding peaks had been recorded within a period of less than two seconds during the scan were used, so as to avoid the effects of fluctuations in sample concentration.

These limitations resulted in 154 possible peak intensity ratios. Some 75 scans were recorded for each isomer, each ratio computed for each scan; the ratios averaged and the corresponding standard deviations calculated. Those ratios were then selected which could distinguish the six isomers by

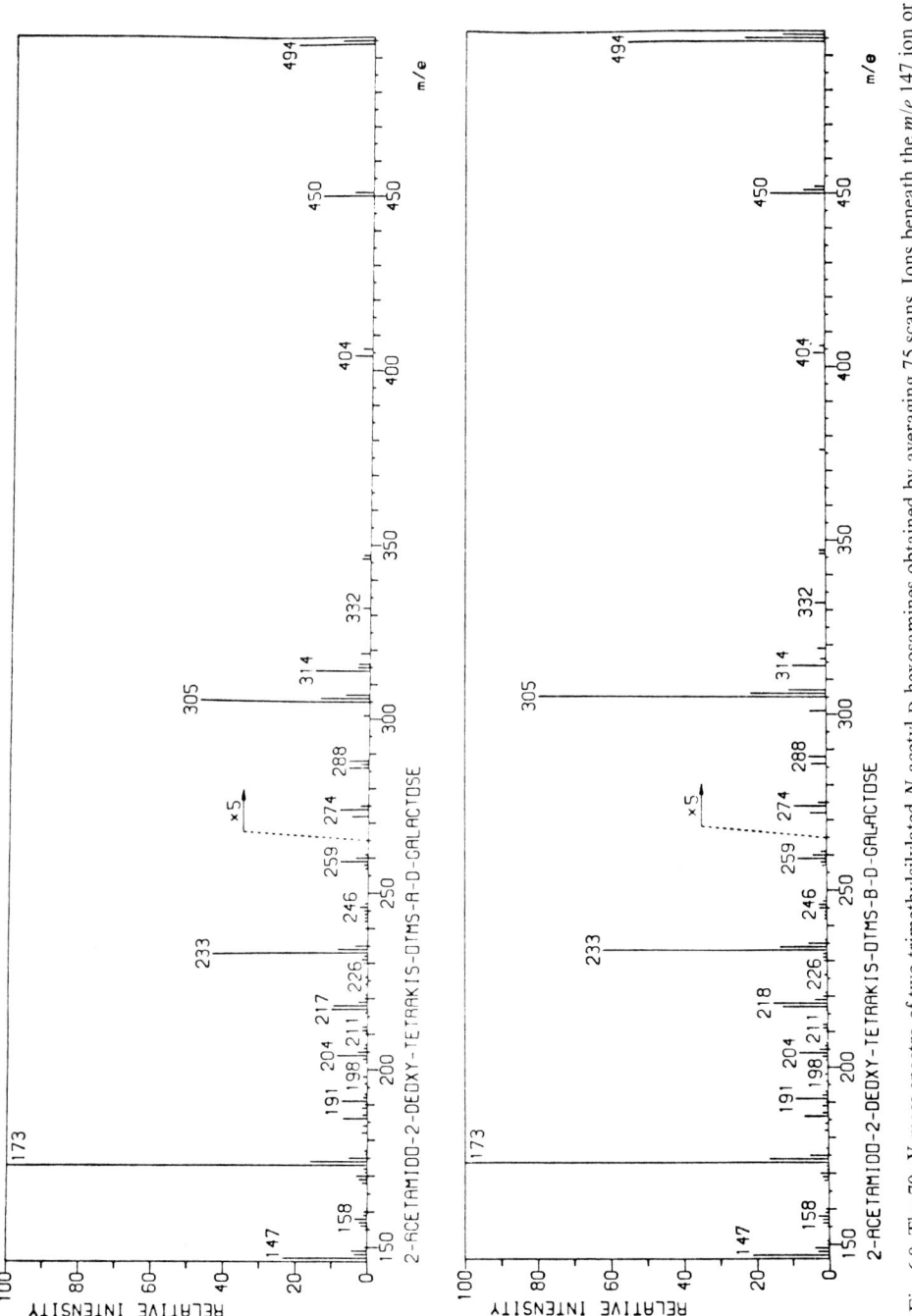

Fig. 6.8. The 70 eV mass spectra of two trimethylsilylated N-acetyl-D-hexosamines obtained by averaging 75 scans. Ions beneath the m/e 147 ion or with abundances less than 0.3% relative to the base peak are omitted. (From ref. 30.)

successive dichotomies. Thus, the first step was to differentiate the N-acetylgalactosamines and the other four stereoisomers. The next steps were to distinguish the two anomers of galactosamine from each other, the glucosamines from the mannosamines, the two anomers of glucosamine from each other and finally the two mannosamine anomers from each other. The Fisher ratio (22) given by $(m_1 - m_2)^2/(\sigma_1^2 + \sigma_2^2)$, where m_1 and m_2 are the means and σ_1^2 and σ_2^2 the variances of the ratio for the two classes, was calculated for each ratio and used as a measure of the separating ability of the ratio for the classes concerned. Seven characteristic ratios were finally chosen by this means (see Table 6.10).

TABLE 6.10

The most characteristic peak intensity ratios[a]

	Trimethylsilylated N-acetyl derivatives of										
	Glucosamine				Galactosamine				Mannosamine			
	α		β		α		β		α		β	
Ratio	mean	s.d.	mean	s.d.	mean	s.d.	mean	s.d.	mean	s.d.	mean	s.d.
I494/I450	3·23	0·68	2·94	0·28	1·66	0·36	3·77	0·67	2·13	0·45	3·18	0·89
I404/I316	1·22	0·10	0·42	0·10	1·84	0·86	2·37	0·86	0·44	0·11	0·81	0·19
I314/I305	1·19	0·23	0·26	0·04	0·33	0·08	0·12	0·04	0·78	0·17	0·79	0·19
I259/I233	2·16	0·47	1·30	0·08	0·18	0·02	0·13	0·02	2·03	0·36	2·18	0·33
I246/I186	0·70	0·05	0·56	0·08	0·33	0·05	0·35	0·03	0·67	0·07	0·58	0·07
I226/I211	2·26	0·22	2·59	0·34	0·49	0·15	0·72	0·13	1·43	0·18	1·63	0·26
I218/I217	0·48	0·07	0·58	0·02	0·96	0·12	1·22	0·19	0·50	0·08	0·48	0·09

[a] Their mean values and corresponding standard deviations (from 75 determinations) from the spectra of six stereoisomeric N-acetylhexosamine trimethylsilyl ethers. (From ref. 30.)

For each reference compound, i, a linear discriminant function g_i is derived:

$$g_i(X) = X^t \Sigma^{-1} M_i + (\log p_i - \tfrac{1}{2} M_i^t \Sigma^{-1} M_i). \qquad (6.12)$$

where X and M_i are the column vector notations for the pattern of the unknown and the mean M for class i, respectively, Σ^{-1} is the inverse of the covariance matrix and p_i is the *a priori* probability of class i. The character t denotes the transpose of the vector. This function is an optimum classifier for a normal distribution of the observed ratio values and assumes equal covariance matrices for each class (11). Equation 6.12 can be rewritten in the form previously used for a linear discriminant function (p. 157), viz:

$$g(x) = w_1 x_1 + \ldots + w_d x_d + w_{d+1}. \tag{6.13}$$

Weight factors for these discriminant functions were calculated according to eqn. 6.12 using the values of the seven characteristic ratios from each reference spectrum. An equal *a priori* probability of each of the six possible sugars was supposed so that p_i in eqn. 6.12 was taken to be 1/6. The unknown spectra were then placed in the category giving the largest g_i value. With the exception of the α- and β-mannosamine derivatives which have very similar spectra, the other stereoisomers could be distinguished with a probability of misclassification of less than 8%, using only a single scan of the unknown, not taken from the training set. The results of the classification of test spectra are shown in Table 6.11. "No classification possible" indicates that the two largest discriminant scores differed by less than 1% of the largest score.

TABLE 6.11

Classification results on test spectra[a]

TMS-N-acetyl derivative	Number of spectra recorded	Number of misclassifications	Misclassified as	No classification possible
α-D-glucosamine	73	0	—	9
β-D-glucosamine	59	4	α-D-mannosamine	12
α-D-galactosamine	53	4	β-D-galactosamine	0
β-D-galactosamine	28	1	α-D-galactosamine	0
α-D-mannosamine	46	2	β-D-mannosamine	8
β-D-mannosamine	57	25	α-D-mannosamine	23

[a] (From ref. 30.)

6.6 UNSUPERVISED LEARNING—CLUSTER ANALYSIS

In pattern recognition, if the sought-for property is not exactly known or if examples of the separate classes are non-existent, then the application is one of unsupervised learning (1, 19, 51). The majority of unsupervised learning methods are actually cluster analysis methods using a number of different criteria to locate "natural" clusters in data points corresponding to "natural" classes in the data.

An example of a cluster analysis approach is a graph theoretical method known as the shortest spanning path (SSP) used by Heller and co-workers

(52) in an attempt to categorize the spectra of compounds containing only one sulphur atom. This procedure creates an ordered list of data points which reflects the shortest path through these points. The process may be represented as in Fig. 6.9 where the data points (A) have been connected by a shortest spanning path whose distance weights are marked on each edge (B).

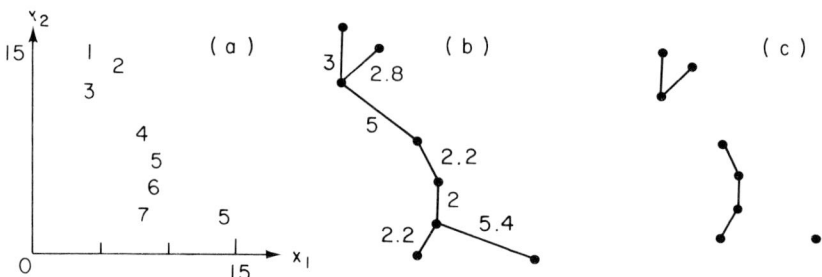

Fig. 6.9. Cluster analysis using the shortest spanning path method. For details, see text. (Adapted from ref. 53.)

Many intuitive means of forming clusters from the SSP graph are available. For example, an edge with a large weight relative to the other branches from its nodes may be broken to form a cluster, as in Fig. 6.9(c). In the case of the sulphur spectra, the re-ordered list from the SSP procedure was divided into segments using the intuitive judgement of a chemist. Using this method, some success was achieved in the recognition of classes of compounds that could be distinguished by their mass spectra.

Another approach to cluster analysis is through the use of display techniques such as non-linear mapping (NLM) (1, 22, 53–5). Non-linear mapping techniques attempt to map the original n-dimensional pattern data onto a two-dimensional form according to various criteria such as maintenance of interpattern distances. The use of such a two-dimensional plot, which corresponds as closely as possible with the original configuration in n-dimensional space, allows an operator to make visual judgements on clustering and classification of unknowns in the data.

This technique has been used in the classification of bacteria by the analysis of the mass spectra of the pyrolysis products from whole bcateria (55). Figure 6.10 shows typical low voltage mass pyrograms of two strains of Listeria bacteria, representing two different serotypes, viz. type I and type IVb. A few differences, e.g. at m/e 82, 110 and 128, are visible in these spectra. However, a non-linear map of the 40 most characteristic mass peaks from this data, selected by a standard weighting procedure, shows two well defined clusters for the two serotypes (Fig. 6.11).

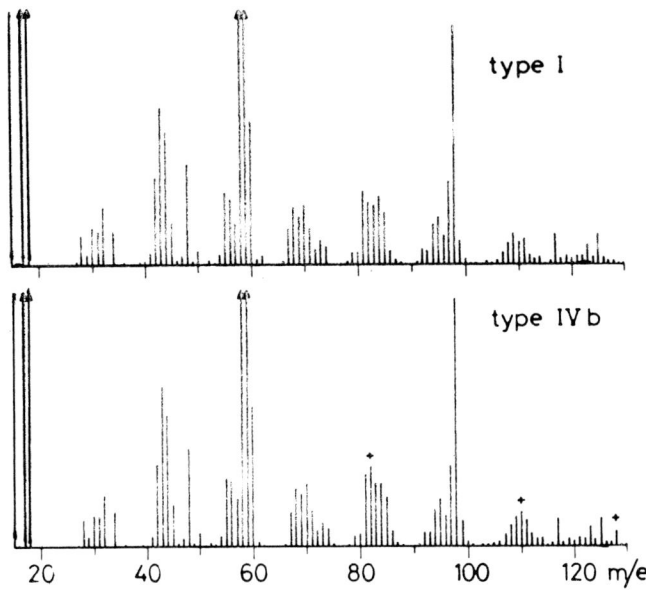

Fig. 6.10. Low voltage pyrolysis mass spectra of two strains of Listeria bacteria. Averaged data. (From ref. 55.)

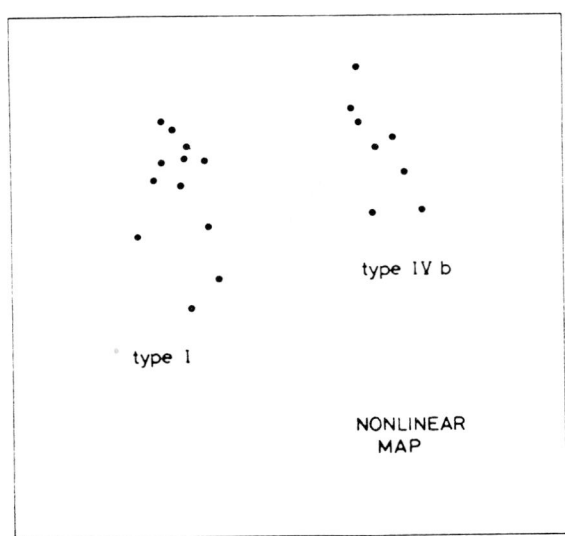

Fig. 6.11. Nonlinear map of characteristic low voltage pyrolysis mass spectral data from 20 Listeria strains. Each point represents an average of 3–4 analyses/strain. (From ref. 55).)

NLM has also been used by Ting *et al.* (22) to explore relationships between mass spectra and the pharmacological activity of drugs. In this case, however, mapping of the original data was not particularly successful as only a moderate degree of clustering was observed. More distinct clustering of this data could be observed by mapping either after weighting or after applying feature reduction methods to the data.

Cluster analysis techniques have also been applied to the identification of the source of archaeological material using mass spectrometric data. Soapstone fragments, excavated from a site in North Germany, were analysed for trace elements by spark source mass spectrometry. Cluster analysis of this data demonstrated the existence of five types of soapstone material on the site and a parallel analysis of samples from Norwegian soapstone quarries showed that one of the types found on the site originated from a newly discovered Viking quarry in South Norway (56).

6.7 CLASSIFICATION USING FACTOR ANALYSIS

An important general transformation that can lead to the classification of data by establishing the optimum number of categories and their class definitions forms the basis of factor analysis (57) (cf. pp. 90–91). Classification of mass spectra using factor analysis requires the transformation of the original data matrix into a covariance or correlation matrix which is then analysed to yield independent factors that are linear combinations of these correlations. Successive factors account for the greatest possible residual variance and these factors are extracted until the residual variance reaches a level corresponding to experimental error. The number of factors extracted represents the number of independent variables required to describe the set of spectra.

The factors initially extracted in this analysis are referred to as principal factors. While principal factors are mathematically simple, they are not directly related to the experimental data, nor are they a unique set. Other factors, e.g. typical factors and partial factors (58), are more directly related to the experimental data. Thus, in the analysis of the mass spectra of the 22 possible $C_{10}H_{14}$ isomers (58), three principal factors were found sufficient to describe the data. It was then possible to select by factor analysis techniques, the three isomers and the three masses which best typify the factors of the data. The three isomers and three masses both form sets of typical factors. On the other hand, partial factors may be exemplified by a set of partial mass spectra which best typify the factors of the data. Yet another type of factor is a basic factor, where an attempt is made to use known properties, such as the ionization potential of a compound as variables. However, attempts to describe mass spectra in this way have so far met with little success (58).

Partial factors seem to provide the best basis for classification. Thus, by using various plotting techniques, Rozett and Petersen (59) were able to demonstrate the separation of the 22 $C_{10}H_{14}$ isomers into five distinct classes, each described by one or more of the partial factors (see Table 6.12 and Fig. 6.12).

TABLE 6.12

The 22 isomers of $C_{10}H_{14}$[a]

1	n-Butylbenzene
2	Isobutylbenzene
3	sec-Butylbenzene
4	tert-Butylbenzene
5	1-Methyl-2-propylbenzene
6	1-Methyl-3-propylbenzene
7	1-Methyl-4-propylbenzene
8	1,2-Diethylbenzene
9	1,3-Diethylbenzene
10	1,4-Diethylbenzene
11	1-Isopropyl-2-methylbenzene
12	1-Isopropyl-3-methylbenzene
13	1-Isopropyl-4-methylbenzene
14	1,2-Dimethyl-3-ethylbenzene
15	1,2-Dimethyl-4-ethylbenzene
16	1,3-Dimethyl-2-ethylbenzene
17	1,3-Dimethyl-4-ethylbenzene
18	1,3-Dimethyl-5-ethylbenzene
19	1,4-Dimethyl-2-ethylbenzene
20	1,2,3,4-Tetramethylbenzene
21	1,2,3,5-Tetramethylbenzene
22	1,2,4,5-Tetramethylbenzene

[a] (With permission from ref. 59. Copyright from American Chemical Society.)

The corner clusters (Fig. 6.12) are primary clusters, each fairly closely identified with the partial factor. The edge clusters, isomers 8–10 and isomer 4 are linear combinations of the corner properties. Other factor analysis results can suggest an identity for each of the corner properties into which the mass spectra of the isomers have been resolved. The partial factor of corner A is a fragmentation pattern associated with $C_9H_{11}^+$ and similar ions. It shows the property of isomers which lose a single carbon atom from the parent ion. Similarly, corners B and C represent the loss of two (to give $C_8H_9^+$ primarily) and three carbon atoms (to give $C_7H_7^+$ primarily).

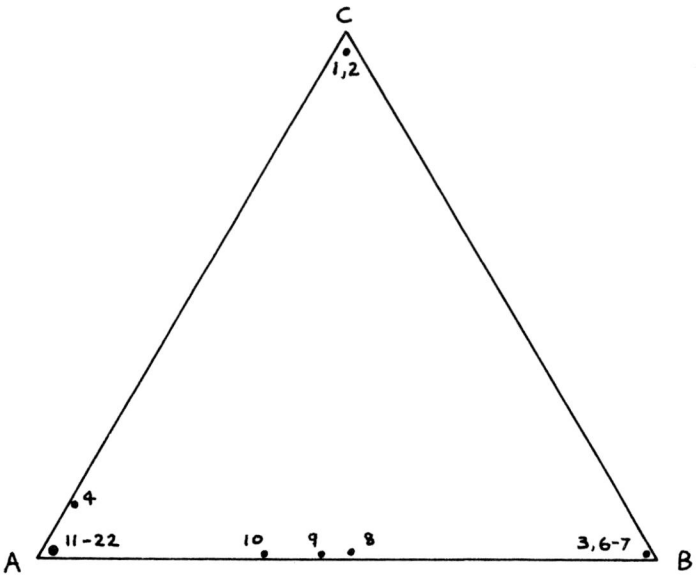

Fig. 6.12. Triangular plot of partial factor analysis data from 22 $C_{10}H_{14}$ isomers. Reference partial factors identified as A, B and C. (Adapted with permission from ref. 59. Copyright from American Chemical Society.)

REFERENCES

1. Kowalski B. R. and Bender C. F. (1973). *J. Amer. chem. Soc.* **95**, 686.
2. Crawford L. R. and Morrison J. D. (1968). *Analyt. Chem.* **40**, 1469.
3. Mathews R. J. (1975). *Int. J. Mass Spectrom. Ion Phys.* **17**, 217.
4. (a) Smith D. H. and Eglinton G. (1972). *Nature*, **235**, 325.
   (b) Smith D. H. (1972). *Analyt. Chem.* **44**, 536.
5. Varmuza K., Rotter H. and Krenmayr P. (1974). *Chromatographia*, **7**, 522.
6. Jurs P. C., Kowalski B. R. and Isenhour T. L. (1969). *Analyt. Chem.* **41**, 21.
7. Gray N. A. B. (1976). *Analyt. Chem.* **48**, 2265.
8. Highleyman W. H. (1962). *Bell Syst. tech. J.* **41**, 723.
9. Rotter H. and Varmuza K. (1975). *Org. Mass Spectrom.* **10**, 874.
10. Varmuza K. (1974). *Monats. für Chemie*, **105**, 1.
11. Nilsson N. J. (1965). "Learning Machines". McGraw-Hill, New York.
12. Kent P. and Gaumann T. (1975). *Helv. Chim. Acta*, **58**, 787.
13. Abe H. and Jurs P. C. (1975). *Analyt. Chem.* **47**, 1829.
14. Jurs P. C. (1971). *Appl. Spectrosc.* **25**, 483.
15. Bender C. F., Shepherd H. D. and Kowalski B. R. (1973). *Analyt. Chem.* **45**, 617.
16. Bender C. F. and Kowalski B. R. (1973). *Analyt. Chem.* **45**, 591.

17. (a) Wangen L. E., Frew N. M., Isenhour T. L. and Jurs P. C. (1971). *Appl. Spectrosc.* **25**, 203.
    (b) Jurs P. C. (1971). *Analyt. Chem.* **43**, 1812.
18. Kowalski B. R. and Bender C. F. (1973). *Analyt. Chem.* **45**, 2234.
19. Kowalski B. R. and Bender C. F. (1972). *J. Am. chem. Soc.* **94**, 5632.
20. Jurs P. C., Kowalski B. R., Isenhour T. L. and Reilley C. N. (1969). *Analyt. Chem.* **41**, 1949.
21. Tunnicliff D. D. and Wadsworth P. A. (1973). *Analyt. Chem.* **45**, 12.
22. Ting K-L. H., Lee R. C. T., Milne G. W. A., Shapiro M. and Guarino A. M. (1973). *Science* **180**, 417.
23. Kowalski B. R., Jurs P. C., Isenhour T. L. and Reilley C. N. (1969). *Analyt. Chem.* **41**, 695.
24. Jurs P. C. (1970). *Analyt. Chem.* **42**, 1633.
25. Zander G. S., Stuper A. J. and Jurs P. C. (1975). *Analyt. Chem.* **47**, 1085.
26. Mathews R. J. (1973). *Aust. J. Chem.* **26**, 1955.
27. Raznikov V. V. and Talroze V. L. (1966). *Dokl. Akad. Nauk SSSR*, **170**, 597.
28. Pichler M. A. and Peronne S. P. (1974). *Analyt. Chem.* **46**, 1790.
29. Burgard D. R., Perone S. P., Wiebers J. L. (1977). *Biochem.* **16**, 1051.
30. Vink J., Heerma W., Kamerling J. P. and Vliegenthart J. F. G. (1974). *Org. Mass Spectrom.* **9**, 536.
31. Gillis R. G. (1971). *Org. Mass Spectrom.* **5**, 79.
32. Jurs P. C., Kowalski B. R., Isenhour T. L. and Reilley C. N. (1969). *Analyt. Chem.* **41**, 690.
33. Isenhour T. L. and Jurs P. C. (1971). *Analyt. Chem.* **43** (9), 20A.
34. Kowalski B. R. and Bender C. F. (1972). *Analyt. Chem.* **44**, 1405.
35. Jurs P. C. (1971). *Analyt. Chem.* **43**, 22.
36. Wangen L. E., Frew N. M. and Isenhour T. L. (1971). *Analyt. Chem.* **43**, 845.
37. Jurs P. C., Kowalski B. R., Isenhour T. L. and Reilley C. N. (1970). *Analyt. Chem.* **42**, 1387.
38. Felty W. L. and Jurs P. C. (1973). *Analyt. Chem.* **45**, 885.
39. Bender C. F. and Kowalski B. R. (1974). *Analyt. Chem.* **46**, 294.
40. (a) Stonham T. J., Aleksander I., Camp M., Shaw M. A. and Pike W. T. (1973). *Electron. Letts* **9**, 391.
    (b) Stonham T. J., Aleksander I., Camp M., Pike W. T. and Shaw M. A. (1975). *Analyt. Chem.* **47**, 1817.
41. Stonham T. J. and Aleksander I. (1974). *Electron. Letts* **10**, 301.
42. Varmuza K. and Krenmayr P. (1974). *Z. Analyt. Chem.* **271**, 22.
43. (a) Schechter J. and Jurs P. C. (1973). *Appl. Spectrosc.* **27**, 30.
    (b) Zander G. S. and Jurs P. C. (1975). *Analyt. Chem.* **47**, 1562.
44. Clerc J. T., Naegeli P. and Seibl J. (1973). *Chimia*, **27**, 639.
45. Varmuza K. (1974). *Z. Analyt. Chem.* **268**, 352.
46. Varmuza K., Krenmayr P. (1973). *Z. Analyt. Chem.* **266**, 274.
47. Justice J. B. and Isenhour T. L. (1974). *Analyt. Chem.* **46**, 223.
48. Gray N. A. B. (1977). (Personal communication.)
49. Cover T. M. (1968). *I.E.E.E. Trans. on Info. Theory*, **IT-14**, 50.
50. (a) Franzen J. (1974). *Chromatographia* **7**, 518.
    (b) Franzen J. and Hillig H. (1974). *In* "Advances in Mass Spectrometry" (Ed. West A. R.). Vol. 6, p. 991. Applied Science Publishers, Barking.
51. Kowalski B. R. (1974). "Computers in Chemical and Biochemical Research" (Eds, Klopfenstein C. E. and Wilkins C. L.) Vol. 2, Ch. 1. Academic Press, New York.

52. Heller S. R., Chang C. L. and Chu K. C. (1974). *Analyt. Chem.* **46**, 951.
53. Meisel W. S. (1972). "Computer-Oriented Approaches to Pattern Recognition". Academic Press, New York.
54. Chu K. C. (1974). *Analyt. Chem.* **46**, 1181.
55. Meuzelaar H. L. C., Kistemaker P. G., Eshuis W. and Boerboom H. A. J. *In* "Advances in Mass Spectrometry", Vol. 7. (In press.)
56. *Europ. Spectrosc. News* 1975, **1**, 19.
57. Rozett R. W. and McLaughlin Petersen E. (1975). *Analyt. Chem.* **47**, 1301.
58. Rozett R. W. and McLaughlin Petersen E. (1975). *Analyt. Chem.* **47**, 2377.
59. Rozett R. W. and McLaughlin Petersen E. (1976). *Analyt. Chem.* **48**, 817.
60. Justice J. B., Anderson D. N., Isenhour T. L., Marshall, J. C. (1972). *Analyt. Chem.* **44**, 2087.
61. Vink J., Bruins Slot J. H. W., de Ridder J. J., Kamerling J. P. and Vliegenthat J. F. G. (1972) *J. Am. chem. Soc.* **94**, 2542.

7

Spectrum Interpretation

7.1 INTRODUCTION

In this chapter, we discuss the elucidation of the structure of unknowns using fragmentation rules and we see how these methods can be applied using data systems. The basis for these methods has so far been knowledge of fragmentation patterns appropriate to various compound types and substructures accumulated by mass spectroscopists. That knowledge, although imperfect, is considerable (44), but cannot always easily be formalized into rules for use in interpretative programs. The knowledge will grow however, especially as more comprehensive collections of mass spectral data become available and it is likely that the use of the computer-based techniques that associate known structural and spectral features will also improve the formalization of this knowledge (see Section 7.5.4).

7.2 INTERPRETATION OF LOW RESOLUTION SPECTRA

Pettersson and Ryhage (1, 2) were among the first to employ the interpretative approach together with a data system, although similar interpretative schemes used without a data system had been described much earlier (39). The programs described by Pettersson and Ryhage were particularly designed to analyse the low resolution spectra of saturated hydrocarbons (1) and fatty acid methyl esters (2). The first stage of the procedure involved a coarse selection based on the calculation of an ion series spectrum for the unknown followed by a fine selection based on individual ion intensities. This overall selection procedure was used to assign the unknown to one of the above mentioned categories or to decide that it belonged to some other class for which no further subroutine was available. Existing subroutines were then used to elicit more detailed structural information for unknown compounds of the appropriate class using the intensities of characteristic peaks in their spectra. For example, the subroutine for saturated hydrocarbons was able to completely identify saturated straight chain and monomethyl substituted hydrocarbons from C_6 to C_{30}.

Similar subroutines, incorporated into a more extensive interpretative program for the analysis of low resolution mass spectra, were developed by Crawford and Morrison (3). The first stage of their program assesses the probability of the compound belonging to one of the 12 following classes: aromatic, ester, ether, acid, ketone, aldehyde, alkene, alkane, alcohol, cycloalkane, diene or amine. This is accomplished by a calculation of the similarity of the unknown spectrum to those of known members of these classes using as criteria the four most intense peaks, four most significant peaks, ion series spectra and distance from the average spectrum for the class (cf. Section 6.2). Knowledge of the class to which a compound belongs, together with a nominal mol. wt estimated by the program, is used to try to limit the possible atomic formulae for the compound as much as possible.

The program to estimate nominal mol. wt examines the highest mass significant peak (significant = background + 20%) and then compares the intensity of successive peaks one mass unit lower until a peak is reached which appears not to be an isotope peak. This is taken to be the molecular ion and is confirmed by a number of consistency checks. The first check is for consistency with the molecular class and the second is for the absence of peaks corresponding to improbable losses from the molecular ion. A further check, based on the greater stability of even electron ions, is made such that if the molecular mass is even, the sum of the odd mass fragment intensities is greater than that of the even and vice versa. Following these checks, the appropriate specialized program for the class, where it exists, is entered. For some classes, no such specialized program is available and in this case, a general interrogation routine based on the presence of diagnostic masses and losses in the unknown spectrum is used. In either case, all structural units identified are then used to try to build a complete structure using a special structural code devised for this purpose (4). A structure drawing program is also available as a final stage.

The specialized subroutines for amines and alcohols are described in more detail in a paper by O'Brien and Morrison (5). These subroutines take the form of a branched questionnaire in which a systematic interrogation of the low resolution spectrum is made. The final output of the program is the type of amine or alcohol, i.e. primary, secondary or tertiary, and the size of the various substituting alkyl groups. Extensive tests of the alcohol subroutine were also carried out in which the data was modified to simulate (a) the substantial variation of the peak intensities in fast scanned spectra and (b) the effect of mass discrimination caused by the recording of spectra on different instruments.

In connection with the method of recognition of the molecular ion used by

Crawford and Morrison, it is of interest to note an entirely different approach to the deduction of the mol. wt when a molecular ion is not present, adopted by Jardine (6). Using a complete low resolution spectrum, all possible sums of the masses taken two by two are computed to give a series of possible values for the mol. wt. Each mass produced in this way has an associated probability value proportional to the summed intensities at that mass. The search for the molecular ion is then based on the argument that maximum probability should coincide with the mol. wt of the compound being studied. Whilst the method was not uniformly successful, at least half 50 compounds tried, gave a result which was either the mol. wt or the mol. wt less one.

A computer routine designed to locate potential molecular ions of trimethylsilylated steroids has been used by Sjövall and coworkers (7). The routine is able to compute a score based on the presence of a potential molecular ion and associated fragment ions and provides this score, together with the mol. wt and a suggested structure as output. Other methods for the deduction of mol. wt are described on p. 195 (using high resolution data) and on p. 204 (using low resolution data).

A system for the classification of compounds from their nominal mass spectra has been described by Gray and coworkers (8, 9, 10). The classifications used are branching tree schemes (e.g. Fig. 7.1). Each classification

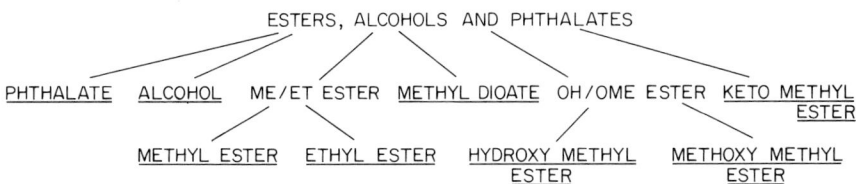

Fig. 7.1. A simple hierarchy of chemical classes used in a classification scheme. Spectral data must be provided for representative compounds in each of the underlined terminal classes. (From ref. 10.)

scheme is encoded as an array of integers. Each class is represented in the array by its name and a set of tests for the spectral features necessary for a spectrum to be considered a possible example of the class. The tests are of the type conventionally used by the mass spectroscopist in interpretation, namely the presence of key ions and losses from the molecular ion (11).

Each set of tests is represented by the identification number for the test subroutine followed by the values for any parameters needed by that routine. For example, a requirement for an ion at m/e 73 greater than 20% may be represented by a call to routine 8 with arguments 73 and 20 and is encoded by successive entries in the array of 8, 73 and 20. Each subroutine sets an index register to point to the next array entry following its set of arguments. Sufficient data to provide for up to approx. 70 classes can be stored in $2K$ of core. The program also attempts to predict a mol. wt and gives an indication of the percentage of the total ion current (TIC) used in making a classification (Table 7.1). This latter feature is useful when more than one class is suggested for a particular spectrum; the suggested classification which accounts for the greatest percentage of the TIC in the spectrum is the preferred class (e.g. peak 3 in Table 7.1). This system has been used in the gas chromatography-mass spectrometry (GCMS) analysis of environmental and geochemical samples where, in many cases, these simple classifications provide an adequate analysis. In other applications, a failure to find such classifications could serve to pick out spectra that need further processing by more sophisticated interpretative or library search techniques.

TABLE 7.1

Summarized assignments in a methylated normal fatty acid fraction from a freshwater sediment[a]

		Computer assignments		
Peak	Component	Code name	Mol. wt	Used ions, % TIC
1	Methyl tetradecanoate	NORMAL ME ESTER	242	79
2	Methyl pentadecanoate	NORMAL ME ESTER	256	73
3	Methyl hexadecenoate	{ ME ENOATE	268	43
		{ DIME DIESTER	...	35
4	Methyl hexadecanoate	NORMAL ME ESTER	270	62
5	Gamma lactone	GAMMA-LACTONE	...	43
6	Methyl octadecenoate	ME ENOATE	296	39
7	Methyl octadecanoate	NORMAL ME ESTER	293	66
8	Dimethyl hexadecane-1,16-dioate	DIME DIESTER	314	43
9	Methyl eicosanoate	NORMAL ME ESTER	326	67
10	Methyl heneicosanoate	NORMAL ME ESTER	340	70
11	Dimethyl octadecane-1,18-dioate	DIME DIESTER	344	41
12	Methyl docosanoate	NORMAL ME ESTER	354	64
13	Methyl tricosanoate	NORMAL ME ESTER	368	65
14	Dimethyl eicosane-1,20-dioate	DIME DIESTER	370	45
15	Methyl tetracosanoate	NORMAL ME ESTER	382	61
16	Methyl pentacosanoate	NORMAL ME ESTER	396	65
17	Dimethyl docosane-1,22-dioate	DIME DIESTER	398	33
18	Methyl hexacosanoate	NORMAL ME ESTER	410	59
19	Methyl heptacosanoate	NORMAL ME ESTER	424	62

[a] For corresponding total ion current (TIC) trace, see Fig. 7.2. (Reprinted with permission from ref. 8. Copyright by American Chemical Society.)

SPECTRUM INTERPRETATION

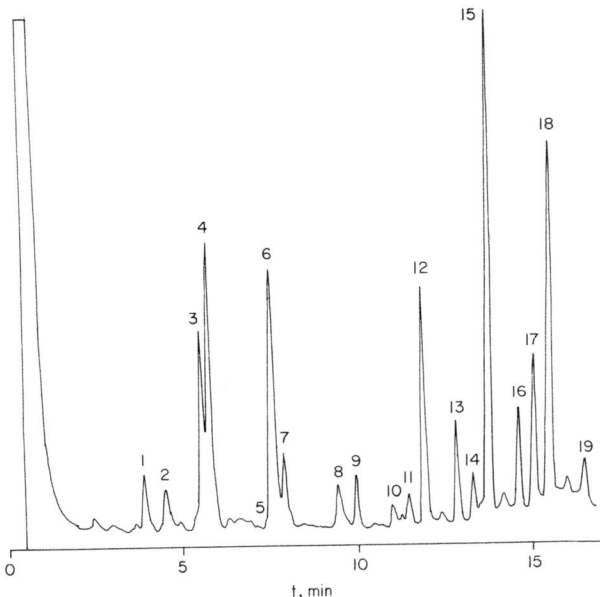

Fig. 7.2. Total ion current trace of a methylated normal fatty acid fraction from a freshwater sediment. Analysis details in Table 7.1. (Reprinted with permission from ref. 8. Copyright by the American Chemical Society.)

Initially, Gray and coworkers used an interactive system to enable the tests for each class to be defined by the operator in terms of a set of standard operations such as a search for a possible molecular ion or series of fragment ions (8). In a later publication (10) the system was improved so that these tests were devised automatically by the program from nominal mass spectra, typical of each terminal class, i.e. classes with no sub-classes (underlined in Fig. 7.1). In a preliminary step, the program causes short spectrum summaries using descriptors such as molecular ion, key fragment ions and losses from the molecular ion, to be abstracted from each typical spectrum. A general description of the spectral features associated with each class is then derived from these individual summaries (e.g. Fig. 7.3).

The new main program also generates a detailed classification scheme by devising tests for these characteristic features. Although this automatic method is capable of handling simple examples, it is more limited than a good spectroscopist in devising spectrum recognition rules. For more complex tests to be devised, the use of a model of the fragmentation processes in order to identify structurally significant fragments would be necessary.

Methoxy methyl esters

Example spectra used to define the class:
  Methyl 11-methoxyoctadecanoate
  Methyl 4-methoxyoctadecanoate
Class characterisation generated by the program:
  Key ion at 313 > 0·8%, molecular ion at key +15.
  Key ion in series from 143 to 241 > 12% with a second key ion occurring at m/e (M—last key +44) and intensity exceeding 50%.
$[M - 31]^+ > 1.5\%$
$[M - 47]^+ > 0.01\%$
$[M - 64]^{+\cdot} > 0.4\%$
Key ion at 45 > 3%
Key ion at 74 > 3%
Series of ions from 55 to 97
Series of ions from 57 to 99

Methyl esters of long chain alkanoic acids

Example spectra used to define the class:
  Methyl tetradecanoate
  Methyl pentadecanoate
  Methyl tridecanoate
  Methyl pentacosanoate
  Methyl tetracosanoate
Class characterisation generated by the program:
  Molecular ion in the series from 228 to 396 with an intensity at least 12%.
$[M - 29]^+ > 0.8\%$
$[M - 31]^+ > 1.5\%$
$[M - 43]^+ > 6\%$
$[M - 57]^+ > 0.8\%$
Key ion at 59 > 3%
Key ion at 74 > 50%
Series of ions from 55 to 153
Series of ions from 57 to 127
Series of ions from 87 to 157

Fig. 7.3. Definitions produced by the main program giving the characteristic features selected by the program for a class formed from two methoxy esters and for a class based on the methyl esters of straight chain alkanoic acids. (From ref. 10.)

7.3 PEPTIDE SEQUENCING

Another early area of application of interpretative routines to the spectra of compounds of a given class was to peptides, using data from the high resolution mass spectra of acyl peptide esters (12, 13, 14). The study of the mass spectra of peptides is of importance because their fragmentation on electron impact gives information which can lead to the determination of the amino acid sequence in the peptides. The cleavage of the peptide chain occurs principally by one of two pathways, either by cleavage at the C—CO bond (type A) or CO—N bond (type B). This simple linear fragmentation was itself an inducement to description using computer techniques.

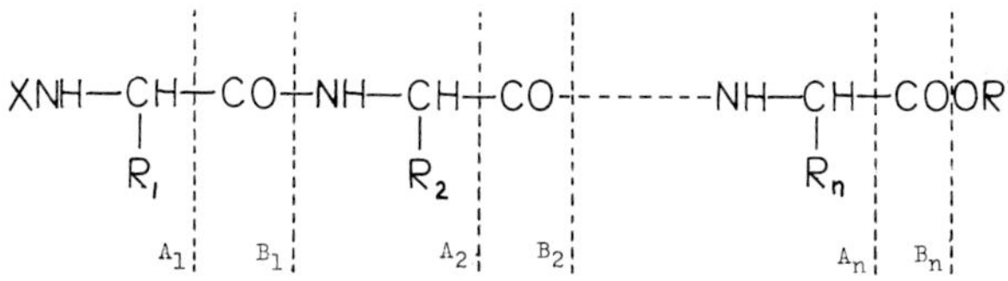

$$XNH-CH{\overset{|}{-}}CO{\overset{|}{-}}NH-CH{\overset{|}{-}}CO{\overset{|}{-}}------NH-CH{\overset{|}{-}}COOR'$$
$$\quad\quad\; R_1 \quad\quad\quad\; R_2 \quad\quad\quad\quad\quad\quad R_n$$
$$\quad\;\; A_1\; B_1 \quad\;\; A_2\; B_2 \quad\quad\quad\quad A_n\; B_n$$

Fig. 7.4. Electron impact induced cleavage of a peptide chain. X is an acyl group.

A typical interpretative scheme was that used by Biemann *et al.* (12). In this, the spectrum was checked for the existence of a peak whose accurate mass corresponded, within the mass measurement error, to fragment A_1, where R_1 was set to correspond to each of the natural amino acids in turn. A similar search was made for fragment B_1 (see Fig. 7.4). Any structures established as candidates for the N-terminal amino acid were then checked by adding fragment A_2 (i.e. —CO—NH—CHR$_2$—), with R_2 now set to correspond to each natural amino acid in turn, to A_1 and searching the spectrum for peaks corresponding to this new dipeptide fragment. At each stage, a check was made to see whether, by addition of the terminal ester group (—CO$_2$R′), a peak corresponding to the molecular ion and therefore indicating the end of the sequence could be found. The program also took into account the abundance of the sequence ions found to try to overcome the problem of alternative sequences arising from the loss of amino acid side chain groups during fragmentation.

Sequencing is now commonly carried out using the more volatile N-permethylated peptides. Programs modified to take account of the simpler fragmentation of these derivatives have been used in the sequencing of peptide mixtures separated by fractional evaporation into the ion source (15, 16).

The preceding computer-based procedures for peptide sequencing use accurate mass data. However, another sequencing procedure developed more recently by Biemann and his coworkers uses only nominal mass data (17). Generation of a mixture of much shorter peptides by random hydrolysis of the original peptide is followed by esterification, acetylation and reduction by LiAlD$_4$ to give a mixture of the corresponding deuterated polyamino-alcohols which are then O-trimethylsilylated (Fig. 7.5). The derivitized hydrolysis mixture is analysed by repetitive scanning GCMS techniques and the location and identification of the trimethylsiyl polyamino alcohols based on three sets of data generated by the computer in this experiment, viz: (a) Mass spectra, showing characteristic C—C bond cleavage (see Fig. 7.5). (b) Mass chromatograms of characteristic ions from these derivatives. (c) Auto-

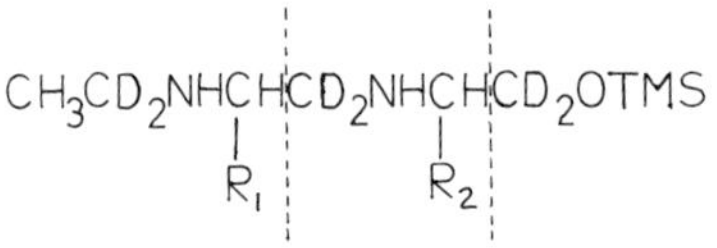

$$CH_3CD_2NHCHCD_2NHCHCD_2OTMS$$
$$\qquad\qquad\;\; R_1 \qquad\;\, R_2$$

Fig. 7.5. Carbon–carbon bond cleavage of a trimethylsilylated deuterated polyaminoalcohol. For details, see text.

matically assigned retention indices (18). In the original publication (17d), this data together with a previous amino acid analysis, was used by the program to reconstruct the 20 amino acid sequence of a fragment of rabbit skeletal muscle actin. Another peptide sequencing method involving random hydrolysis was earlier proposed by Dayhoff and Eck (19), although their computer-based analysis of the hydrolysate was to be carried out using high resolving power accurate mass data obtained from a single scan of the unseparated derivitized hydrolysis products.

7.4 INTERPRETATION OF HIGH RESOLUTION SPECTRA

Apart from peptide sequencing, accurate mass data has been used in the automated structural elucidation of compounds belonging to a fairly limited number of simple structural classes (20–23) such as saturated acyclic ketones, esters, alcohols, amines and ketoalcohols (20) and ketones, amides and amines (21). One general procedure for structural elucidation using accurate mass data is that summarized by Mandelbaum et al. (21), viz; (a) establish the elemental composition and compound type of the unknown and (b) determine the structure of the unknown on the basis of the known fragmentation processes for the specific compound type. In most cases, the peak corresponding to the molecular ion is the one of highest mass in the spectrum with the exception of the corresponding isotope peaks. The exact mass of this peak, taken from the complete high resolution spectrum, will then give the molecular composition of the unknown. There are two situations in which the use of the peak of highest mass is not justified; (a) if the compound fragments so readily that a molecular ion is not observed and (b) if impurities of higher mass are present. Biemann and McMurray (24) developed a computer routine which, on the basis of the full high resolution elemental composition data, was able to reveal some ions due to impurities and was also able to make a reliable suggestion for the molecular ion even when it was absent from the recorded spectrum. In this method, ions present in the spectrum are tested against a number of criteria for a molecular ion. If no ions meet these criteria, further candidate ions are derived by the addition of commonly lost neutral fragments to the ions in the spectrum. A very similar routine was described by McLafferty (20).

According to McLafferty, the principal factors used to select the most probable compound type are ion series data and neutral losses from the parent ion (20). A compound type selected on the basis of this data is checked for consistency with the number of heteroatoms in the molecular ion and its double bond equivalent (DBE). For example, an ester must contain a minimum of two oxygen atoms and one DBE. Both Mandelbaum (21) and McLafferty (20) give details of the subroutines to be used with various functional groups of which that for use with ketones (RCOR') is typical. In this, three types of ions are selected; (A) Ions having the empirical formula $C_nH_{2n}O$, (B) Acyl ions, $C_nH_{2n-1}O$ and (C) Alkyl ions C_nH_{2n+1} (see Fig. 7.6). The heaviest ion of type A is assigned as the molecular ion. The carbon content of all possible pairs of acyl ions is summed and one carbon is subtracted. Possible structural solutions are recorded when this result equals the carbon content of the molecular ion.

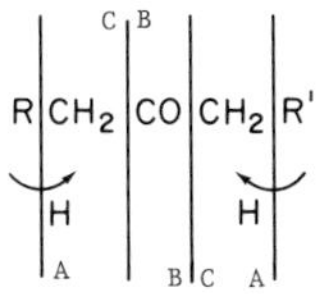

Fig. 7.6. Electron impact induced cleavage of an aliphatic ketone.

A search is then made for the two alkyl ions to be expected from the postulated structures. Their intensities are added to those of the corresponding acyl ions and the structure with the largest accumulated intensity taken as the most probable structure. Ions of type A, which will retain any substituent on the alpha carbon, are now searched for and their intensities added to the accumulated intensity for possible precursor structures. Again, the structure corresponding to the maximum intensity explained is considered to be the most probable one.

A similar but more detailed approach to functional group analysis using high resolution data has recently been described by Hilmer and Taylor (23). Three functional group categories which are expected to have a number of fragmentation patterns in common are defined; viz (a) compounds containing a carbonyl group such as esters, acids, ketones and aldehydes: (b) fully saturated functional groups containing one or more oxygen atoms, such as ethers and alcohols, or containing one or more sulphur atoms, such as thio-ethers and thiols and (c) compounds containing only carbon, hydrogen and the halogens. Once the molecular ion has been identified (using the method of Biemann and McMurray (24)), a thorough analysis of each unknown spectrum is made on the basis of the fragmentation patterns expected for functional groups in each category. The single functional group is chosen that is able to explain the largest percentage of the observed spectral intensity. Using this method, the program was able to recognize the functional group present and correctly explain more than 85% of the observed spectral intensity in most of the samples analysed.

A scheme for structural analysis using accurate mass or nominal mass data that has a similar format to some of the methods used by the Stanford group (see Section 7.5) has been described by Koo and Sedgwick (22). In their scheme, which deals with acyclic amines, alcohols, ketones, aldehydes and ethers, the functional group suspected from a first analysis is classified as mono-, di- or trivalent and substituted by $-CH_3$, $-CH_2-$ or $-CH{<}$ respectively, in separate routines to give equivalent hydrocarbon structures. All possible isomers of the equivalent hydrocarbons are then formed by a structure generator and finally the functional groups suspected substituted back in to give a set of possible structures. Each candidate structure is then subjected to a number of fragmentation processes to produce probable fragment ions. Rearrangement processes specific to the functional group are considered as well as fragmentation of linear carbon chains and breaking the structure at all branch points. This latter process is particularly easy as the Wiswesser line notation used to describe the structures accentuates branching points in the description. Each candidate structure leads to a set of fragment ions which are compared with ions in the observed spectrum and a score based on the coincidence of the ions in the two lists is calculated. Using some 140 mono-functional acyclic compounds as test cases, the correct structure was found in the top three choices in 80% of the cases. The use of elemental compositions from high resolution data was found to be advantageous in obtaining the correct matching and led to more satisfactory scores than did the use of nominal mass spectra.

On the basis of results obtained with this system, Adams and Sedgwick (25) proposed that there was a significant correlation between molecular complexity (expressed as the number of Wiswesser line notation symbols (n)

required to describe a structure) and the prediction rate (P_n) of the system, i.e. the frequency with which the correct structure was chosen as the best fit for such structures (Table 7.2).

$$P_n = (\bar{p})^n. \tag{7.1}$$

In the correlation proposed, (eqn. 7.1), the results for the individual molecular complexity groups are considered as arising from strings of binary decisions with an average predictive ability, $\bar{p}$. However, more data is required to establish the validity of this interesting correlation.

TABLE 7.2

Prediction rate for structures of differing molecular complexity [a]

WLN symbol number n	Best candidate structure		$\% P_n$
	Correct	Not correct	
2	8	2	80
3	14	8	64
4	12	10	55
5	7	6	54
6	1	4	20
7	7	8	46
8	0	1	0

[a] See text for details. (From ref. 25.)

A more flexible system, suitable for unknowns which cannot readily be allocated to classes for which rules have been formalized, was applied by Biemann and coworkers (26) to the interpretation of high resolution spectra. In this, rather than attempting the task of formalizing the body of mass spectral fragmentation knowledge, the computer is allowed to interact through a dialogue with the operator who can then apply intuitive decisions based on experience to the data exhaustively prepared by the computer. A summary and illustration of this approach follows.

First, the elemental compositions of the ions below m/e 100 are calculated to see whether they contain N or O. This can be done exhaustively as the total number of possible compositions below m/e 100 is relatively small. If the compound contains N or O, then a reasonable percentage of the ions below m/e 100 will contain N or O respectively. A further test for the presence of Cl, Br, S or Si is then made by looking for pairs of ions differing by 1.998 ± 0.003 a.m.u. and having intensity ratios between 1:1 and 30:1. The molecular ion is chosen by inspection and possible elemental compositions calculated on the basis of the elements found in the previous two steps together with carbon and hydrogen. Further information from the spectrum may be displayed at this stage in the form of "ion-types" (27). An ion-type refers to the ions belonging to the general series $C_nH_{2n+x}N_yO_z\ldots$ Within each type only n varies while $x, y, z \ldots$ are constant, i.e. it is a homologous series of given double bond equivalent (DBE) and heteroatom content.

Together with ion-type data the computer can also display the result of matching the most intense ions against a reference table of exact mass ions known to be characteristic of certain structural features. The selection of significant information from these displays is a stage at which the operator must then take some decisions based on experience. Once this process has, hopefully, led to the identification of some structural features in the molecule, the computer may be used to search for specific fragment ions which could amplify this structural information and perhaps lead to a complete structure.

In an example given by Biemann (26) of an unknown obtained by hydrolysis of a nucleic acid, the procedure outlined had shown the elemental composition of the unknown to be $C_{15}H_{21}N_5O_4$ (eight DBE) and located key ions representing a derivative of adenine (five carbons, five nitrogens and six DBE) and a pentose unit (five carbons, four oxygens and one DBE). Thus, five carbons and one DBE were unaccounted for. To decide whether these missing atoms were associated with the adenine fragment, it was necessary to search for ions containing five nitrogens and seven DBE. Therefore, a search for ion types $C_yH_{2y+x}N_5$, where $x = -8$ for bond cleavage and $x = -7$ for hydrogen transfer, was made. An ion of the latter type containing all 10 carbon atoms was found in the spectrum, whereas no ion associating the missing five carbon atoms with the pentose unit could be found. Further dialogue, including the introduction of available nuclear magnetic resonance (NMR) data by the operator, finally elucidated the structure as Formula (I):

7.5 HEURISTIC DENDRAL

The heuristic programming approach was initially adopted by the Stanford group as the basis for a general method directed towards the complete interpretation of low resolution mass spectra. Slagle (31) defines a heuristic as "... a rule of thumb, strategy, method, or trick used to improve the efficiency of a system which tries to discover the solutions of complex problems". Heuristic programming is then the programming of a digital computer with an algorithm that uses heuristics. The heuristics for the interpretative program took the form of a simple theory of mass spectral fragmentation processes.

7.5.1 Structure Elucidation

The first application of this approach was to the interpretation of the low resolution mass spectra of ketones (32).

The outline of the approach is shown in Fig. 7.7. From the low resolution mass spectrum and the empirical composition, the program employs a theory of fragmentation processes, stored in the Preliminary Inference Maker (PIM) to decide what functional groups are present within the structure of the unknown. The program now gives the inferred functional groups highest priority, by placing them on GOODLIST. The Structure Generator, using the DENDRAL algorithm (33) then builds all possible candidate molecules within the constraints inferred from the spectrum. Without any such constraints the output generated by the structure generator would be enormous; for example there are 1936 possible acyclic structures corresponding to the composition $C_9H_{18}O$. Thus, the task of the PIM is to produce constraints to reduce such numbers to a more manageable level. The presence of a ketone, for example, is inferred using the same kind of fragmentation rules (now applied to low resolution spectra) as used by Mandelbaum (22) and McLafferty (21) (cf. p. 195). Two significant acyl ions, with masses B and B', are sought that satisfy the relationship $B + B' = M + 28$ (see Fig. 7.8). The corresponding alkyl ions, A and A', must also be found. Even if a ketone is found to be present, the program then checks for the presence of compounds such as aldehydes, ethers or alcohols that would also satisfy the empirical formula. The program also searches for ketone substructures. For example, the presence of a methyl ketone can be inferred from the presence of a peak at m/e 58 (see Fig. 7.9).

Any functional groups or substructures found are placed on GOODLIST

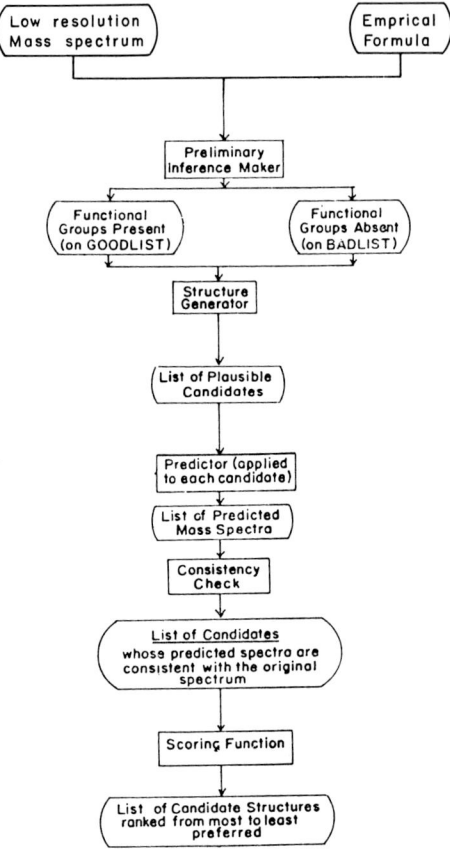

Fig. 7.7. Outline of Dendral approach to spectrum interpretation. (Adapted from ref. 34.)

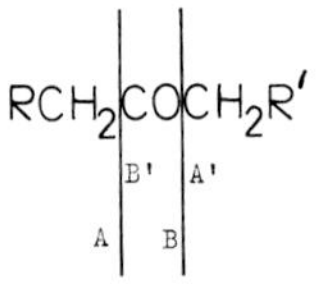

$$RCH_2|CO|CH_2R'$$

Fig. 7.8. Electron impact induced cleavage of an aliphatic ketone.

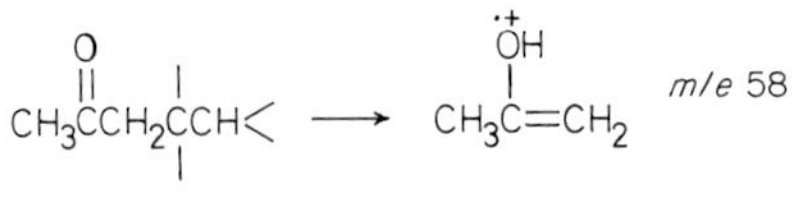

Fig. 7.9. Electron impact induced cleavage of a methyl ketone to give an ion at m/e 58.

and only those full structures then generated that contain the elements on GOODLIST. Any groups or substructures considered to be absent because the appropriate peak is absent are placed on BADLIST and only full structures lacking these elements are generated. For each of the structures produced by the generator, the Predictor subroutine predicts an abbreviated mass spectrum. This is compared with the spectrum of the unknown and inconsistencies between these spectra are found and eliminated. Candidate structures that then remain are finally ranked by the Scoring Function.

After its application to aliphatic ketones, later papers described the use of heuristic DENDRAL in the structural elucidation of aliphatic ethers (34) and aliphatic amines (35). Besides the introduction of NMR data into the program to further reduce the structure generator output, these papers saw the successive expansion of the capabilities of the PIM. For the amines, the PIM produced a complete list of what were termed superatoms, i.e. amine substructures with at least one free valence. The 31 possible superatoms were constructed as all possible combinations of the basic substructures (T, S, P and M) attached to one nitrogen atom (see Fig. 7.10).

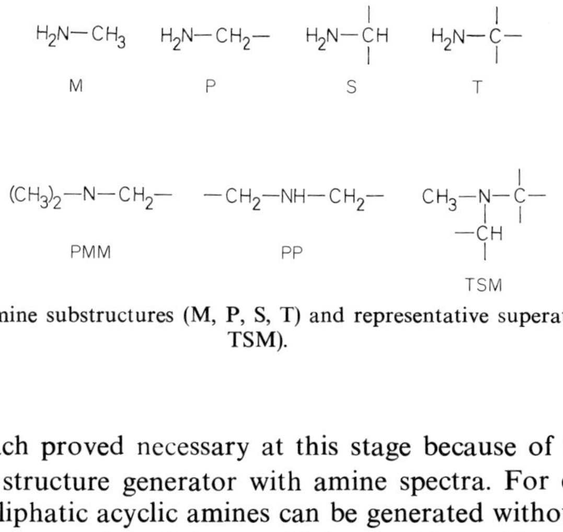

Fig. 7.10. Basic amine substructures (M, P, S, T) and representative superatoms (PMM, PP, TSM).

This approach proved necessary at this stage because of the very large output of the structure generator with amine spectra. For example, 7639 possible C_{12} aliphatic acyclic amines can be generated without constraints. In fact, the efficiency of the Inference Maker in this problem and in a later application to a general set of saturated acyclic monofunctional (SAM) compounds (36) was such that the other two program phases (Structure Generator and Predictor) were not used.

The processes used by the inference maker in dealing with SAM compounds (these have the general formula $C_nH_{2n+v}X$, where X = O, S or N and v = valence of X), may be summarized as follows (36). At the outset, a filter program removes those ions which are relevant to the structural determination of SAM compounds. These ions comprise the α-cleavage series that start from the masses of general formula $CH_2 \cdot XH_{v-1}$ (i.e. 30, 31 and 47 for N, O and S respectively) and the three alkyl ion series C_nH_{2n+1}, C_nH_{2n} and C_nH_{2n-1}. An unknown spectrum is accepted as being that of a SAM compound if no intense ions now remain. Using the α-cleavage ion series, the Inference Maker sums for each heteroatom the total intensity found in the corresponding α-series. The heteroatom with the highest score is taken to be the most likely one present. This choice is also validated using the alkyl ion intensities. Once a heteroatom is chosen, the lowest plausible mol. wt and therefore an empirical formula based on this, is inferred.

Having decided upon the empirical formula, the program then builds up a complete and non-redundant set of superatoms. This set of superatoms contains the heteroatom and every possible set of α-atoms attached to the heteroatom. The program now tries to validate the presence of any of these superatoms from the unknown spectrum. The surviving validated superatoms are next transformed into general molecules by the addition of all remaining carbon atoms in the empirical formula as alkyl groups. Each of these general structures can be tested further against the unknown spectrum and the final output shows those which have passed all such tests requiring the presence of α-cleavage ions, rearrangement ions, alkyl ions, etc. A complete description of these tests may be found in the relevant publication (36). Running the program on 210 unknown spectra always resulted in the correct structure being included in the output.

The next application to structural elucidation in the heuristic Dendral project was to a more complex class of compounds, the estrogenic steroids (37). This work was similar to the work on amines and SAM compounds previously described, in that it did not use a general and systematic program for the generation of all possible molecular structures. The main reason for this was the lack at that time of a formal structure generator that encompassed cyclic structures, such as the estrogenic steroids, as well as acyclic structures. In addition, this work did not use a predictor phase for ranking of final structures. Essentially the program was an extension of the inference making phase of Heuristic DENDRAL now referred to as "Planner". The program was also a departure in that it made use of complete high resolution (accurate mass) data together with any available metastable ion data.

Given the structural skeleton for estrogens (see Fig. 7.11), the program calculated the empirical formula, mass and degree of unsaturation of this skeleton. The estrogen skeleton has the empirical formula $C_{18}H_{24}$ and seven DBEs. Any ion in the spectrum with a higher mass and at least this number of carbon atoms was therefore a potential molecular ion. A candidate molecular ion was discarded if it could be formed by loss of hydrogens from another candidate at higher mass or if it could be shown to be just an isotope of a lower mass peak. The program also eliminated any candidate which showed no metastable transition to any daughter ion in the spectrum. This molecular ion algorithm is perhaps the weakest point of the whole program (see Section 7.5.2) for a more recent molecular ion algorithm employed by the Stanford Group.

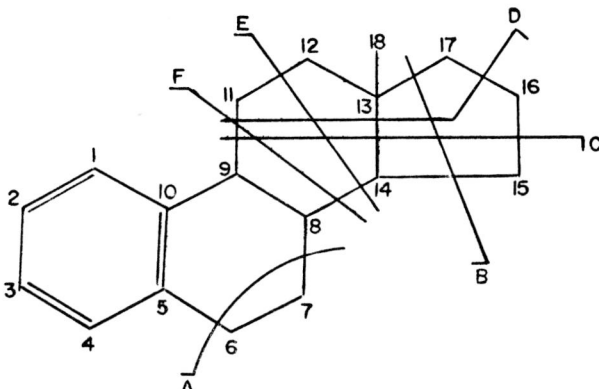

Fig. 7.11. Symbolic representation of the fragmentation rules employed by the program for estrogens. (Reprinted with permission from ref. 37. Copyright by the American Chemical Society.)

The program now seeks to place substituents required to make up the molecular formula on this skeleton. It does this by searching for evidence for the fragmentations B to F with allowed hydrogen transfers which are typical of estrogenic steroids (see Fig. 7.11), considering all possible placements of the required substituents. In general, more than one ion will be found as evidence for a given fragmentation. The program then retains only those substituent placings where the sum of all corresponding ion intensities exceeds a given threshold value.

After this analysis, the program considers the consequences of every combination of substituents, taking one set of substituents from each fragmentation result. As the number of possibilities for each fragmentation increases, the number of combinations increases rapidly. Thus, any rules (heuristics) that reduce the number of possibilities will save considerable computing effect. As an example, in the first pass through the program, only the most intense ions indicated for fragmentation B were used (previous experience had pointed to the high intensity of ions arising by this fragmentation) whereas all ions above a threshold were considered for the other fragmentations. If the first pass gave no structures these restrictions were successively relaxed until structures were obtained.

Finally, the structures deduced were filtered retrospectively by a number of rules that were thought appropriate in particular circumstances. For example, prohibition of chemically implausible combinations of substituents (analogous to BADLIST (Fig. 7.7)), inclusion of substituents known to be present (analagous to GOODLIST (Fig. 7.7)) and rules based on natural properties, e.g. known substitution patterns, in certain cases. Tests of this program using the spectra of 43 estrogenic steroids yielded only six cases where incorrect structures were proposed and only one where no structure at all was proposed.

In the case of two of the compounds which were inadvertent mixtures of two estrogens, two molecular ions were found in each case and the structures of both components were determined by the program run in its normal manner. This ability to directly analyse mixtures was deliberately applied in a later paper (38) to mixtures of estrogens from pregnancy urine. High resolution spectra were recorded as the sample was slowly volatilized from the direct insertion probe into the source. Several spectra representative of the mixture were obtained by this means. In addition a low electron voltage spectrum, which showed only molecular ions and metastable data, which defines the routes by which various ions decompose, were recorded when sample quantities permitted. The low electron voltage data allowed ready location of molecular ions and hence elemental compositions of the components using the high resolution data. The metastable data proved to be a very useful constraint on the number of possible structures produced by the planner, by providing connectivity data between peaks in the mixture spectrum. The program was very successful in analysing the major constituents of microgram quantities of unknown estrogens even in the presence of significant impurities. Intermediate results provided by the planner can be used to suggest further metastable scanning experiments that would provide useful information on the composition of the mixture.

7.5.2 Recognition of the Molecular Ion

The recognition of the molecular ion plays a key role in the methods of structural elucidation using Heuristic DENDRAL discussed so far. To assist in this process the Stanford school developed a method which parallels those of Biemann (24) and McLafferty (20) and which can predict a molecular ion from a spectrum whether a molecular ion is present or not (40). However, unlike these methods, where candidate molecular ions are judged on the basis of being related to lighter ions in the spectrum by one of an arbitrary set of commonly lost neutral fragments, e.g. CH_3, H_2O or CO, a set of "secondary losses" derived solely from the nominal mass spectrum being analysed is used to infer candidate molecular ions. The spectrum is initially corrected for ^{13}C isotope contributions and then transformed to a reduced spectrum containing only the most intense peaks in each mass cluster found in the original spectrum. A cluster is defined as a group of peaks in which successive peaks are separated by less than three mass units. Before generating the secondary loss set, the spectrum is checked for molecular ion candidates by seeing whether there are any "bad losses" (i.e. improbable losses such as 4, 5, 6, 11 or 12 a.m.u.) from a candidate ion to neighbouring clusters and whether there is a peak in the top cluster of the appropriate parity with respect to the nitrogen content.

Next, a list of "secondary losses" is exhaustively generated as the losses between all ions in the reduced spectrum. "Bad losses" are excluded from this list as are impossible composition differences when working with high resolution spectra. In addition these losses are restricted to less than 115 a.m.u. and even then cannot be greater than approximately half of the largest observed mass. These losses, excluding those from any possible molecular ion candidate, are then combined with the list of fragment masses from the lower half of the spectrum to give the complete "secondary losses" set. To generate the molecular ion candidates, each member of the secondary loss set is added to each member of the upper half of the reduced spectrum of the appropriate parity. The secondary loss list can be the complete one referred to above, or a reduced version prepared by arranging the losses in the form of ion series and using only the first two or three lower members of each series. These candidate molecular ions are tested as before by seeing whether there are any improbable losses from each to other ions in the spectrum. If the complete secondary loss list has been used, then candidate molecular ions that are retained are given a ranking based on the intensities of the contributing ions. Results so far indicate that there is a high probability (greater than 95% for 250 spectra of more than 10 classes) that the molecular ion will be in the top five candidates. Typical results are shown in Table 7.3 Of the 250 spectra tested 100 did not show significant molecular ions. This program (Molion) has now been incorporated as part of the program Planner.

7.5.3 Structure Generation

The original concept of Heuristic DENDRAL as a tool for structural elucidation was a scheme with three distinct phases; (a) Structural inference or planning from the spectrum. (b) Generation of all possible structures with constraints provided by the substructures inferred in (a). (c) Testing of generated structures against the original spectral data (PLAN—GENERATE —TEST). As we have seen, because of the size of the output from the structure

TABLE 7.3

Sample set of results showing molecular ion prediction and ranking using program "Molion"[d].

Compound	Mol. formula	Highest mass present	Fragment missing[a]	Mol. wt	Ranked at no.[b]
Ritalin	$C_{14}H_{19}NO_2$	172 (M − 61)	$C_2H_5O_2$	233	4
Pentobarbital	$C_{11}N_{18}N_2O_3$	197 (M − 29)	C_2H_5	226	2
Mebutamate	$C_{10}H_{20}N_2O_4$	175 (M − 57)	C_4H_9	232	3
Tridecan-7-one	$C_{13}H_{26}O$	155 (M − 43)	C_3H_7	198	4
Succinic acid methyl ester	$C_6H_{10}O_4$	116 (M − 30)	CH_2O	146	2
Caprylic acid methyl ester	$C_9H_{18}O_2$	129 (M − 29)	C_2H_5	158	3
Glutaric acid methyl ester	$C_7H_{12}O_4$	129 (M − 31)	CH_3O	160	1
Maleic acid butyl ester	$C_{12}H_{20}O_4$	173 (M − 55)	C_4H_7	228	2
N-TFA α-alanine[c] butyl ester	$C_9H_{14}NO_3F_3$	186 (M − 55)	C_4H_7	241	2
N-TFA norleucine butyl ester	$C_{12}H_{20}NO_3F_3$	227 (M − 56)	C_4H_8	283	2
N-TFA valine butyl ester	$C_{11}H_{18}NO_3F_3$	227 (M − 42)	C_3H_6	269	2
N-TFA threonine butyl ester	$C_{12}H_{15}NO_5F_6$	323 (M − 44)	C_3H_8	367	1
N-TFA phenylalanine butyl ester	$C_{15}H_{18}NO_3F_3$	216 (M − 101)	$C_5H_9O_2$	317	4
n-Undecyl alcohol	$C_{11}H_{24}O$	154 (M − 18)	H_2O	172	1
4-Methyloctan-4-ol	$C_9H_{20}O$	129 (M − 15)	CH_3	144	1

[a] These fragment composition losses from the molecular ion are only postulated. Their validity could only be confirmed by high resolution studies.
[b] Note "ranked at number 1" is the program's best choice for a molecular ion candidate.
[c] TFA refers to the trifluoroacetyl derivative.
[d] Reprinted with permission from ref. 40. Copyright by the American Chemical Society.

generator, the inference making section was expanded considerably to provide maximum constraints, so that in some cases the generator and testing phases were not employed at all (35, 36, 41). In addition, lack of a generator for cyclic structures at that time eliminated this phase from the estrogen analysis (37).

However, in parallel with this work, a completely general structure generator was developed (42) and programs were subsequently written to make it possible to limit its output at any stage by specification of substructures to be included or excluded (43). The structure generator in its most recent form is able to produce a complete list of all structural isomers consistent with a given empirical formula without redundancy.

The program begins by constructing vertex-graphs which are cyclic skeletons from which nodes of degree less than three have been deleted. Thus, vertex graph I is the progenitor of structures II and III amongst many others (see Fig. 7.12).

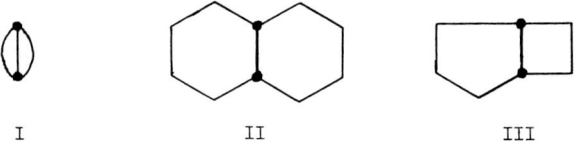

I II III

Fig. 7.12. Illustration of vertex graph I as progenitor of structures II and III. (From ref. 50.)

The vertex graphs from which structures are to be constructed can be specified for a given empirical formula by a series of calculations. In these the formula is first stripped of all hydrogens and these replaced by effective degrees of unsaturation (U). Thus C_6H_8 becomes C_6U_3 and C_3H_6O becomes C_3OU. This "pot" of atoms is then partitioned in all possible ways; each partition consists of those atoms assigned to one or more "superatompots" and a "remaining pot". Each superatompot is a collection of atoms from which all possible ring superatoms can be constructed by further partitions, followed by selection of appropriate vertex graphs from a catalogue maintained by the program and expansion of these graphs. A ring superatom is a ring system with at least one free valence in the completed structure. The atoms in the remaining pot will form the acyclic parts of the final structures when combined in all possible ways with the ring superatoms from the corresponding initial partition. All the possible partitions of C_6U_3 are shown in Table 7.4.

For a full description of the generator program and examples of appropriate outputs, the reader is referred to the original paper. However, as an example of the size of the potential output of the structure generator the

TABLE 7.4

Allowed partitions of C_6U_3 into superatompots and remaining pot[a]

Partition number	Number of super-atompots	Superatompot number			Remaining pot
		1	2	3	
1	1	C_6U_3			
2	1	C_5U_3			C_1
3	1	C_4U_3			C_2
4	1	C_3U_3			C_3
5	2	C_4U_2	C_2U_1		
6	2	C_3U_2	C_2U_1		C_1
7	2	C_2U_2	C_2U_1		C_2
8	2	C_4U_1	C_2U_2		
9	2	C_3U_1	C_2U_2		C_1
10	2	C_3U_2	C_3U_1		
11	3	C_2U_1	C_2U_1	C_2U_1	

[a] (From ref. 42.)

TABLE 7.5

Number of isomers of some selected organic compounds[a]

Formula	Example	Number of structures
C_6H_{14}	Hexane	5
C_6H_{10}	Cyclohexene	77
C_6H_6	Benzene	217
$C_6H_{10}O$	Cyclohexanone	747
C_6H_6O	Phenol	2237

[a] (From ref. 50.)

total number of structures for several empirical formulae is given in Table 7.5. The great increase in the number of structures on introduction of a single heteroatom is particularly noteworthy.

Use of this complete structure generator in structural elucidation has been set in the framework of a program called CONGEN (constrained structure generation (43). In this program, structure generation utilizes an assembly of inferred structural fragments, referred to as "superatoms", (the inference of structure may be from many other sources besides mass spectrometric data) and individual atoms which together make up the elemental composition of the molecule. Intermediate structures are first generated using only the name of the superatoms. Superatom names are subsequently expanded into their full identities and placed in the generated structures by a process called imbedding. Both the generation of intermediate structures and the imbedding process may be constrained by the specification of desired or of undesired structures.

The whole essence of the program is its interactive nature which gives the user access to the problem at the level of intermediate structures. Thus, large numbers of final structures may be removed by discarding only a few intermediate structures. The ability to guide this procedure interactively to a solution helps prevent the generation of unmanageable numbers of undesired structures. Because of the scale of CONGEN and the structure generator program itself, it is not possible here to give even abbreviated examples of their output. For this data, the reader is referred to the original papers (42, 43).

7.5.4 Formalization of Fragmentation Rules

As will be appreciated, all the applications of the DENDRAL program to spectral interpretation have required the prior formulation of relevant fragmentation rules (heuristics), particularly for the planning or inference making phase. Where are these rules to come from? Initially, they represented a summary made manually by mass spectroscopists of the processes they had postulated and found adequate to explain, at least in part, the mass spectra of known compounds. However, two programs have since been developed which attempt to fit, in a systematic manner, formal fragmentation reactions to the spectra of known compounds. In this context, the computer has the advantages of being completely thorough and being able to deal with large amounts of data. These programs are the INTSUM program of the Stanford school (29) and the Ion Generator program of Delfino and Buchs (28).

Information fed to the INTSUM program comprises the basic skeleton of the compound class to be investigated together with known structure–spectrum pairs from this class. The INTSUM program then follows three stages which conform to the usual PLAN—GENERATE—TEST format. The first stage, named ALLBREAKS, generates a non-redundant list of all possible fragmentations of the basic skeleton which produce smaller, unique fragments. This process may be restricted or extended in a number of ways under the operator's control. For example, cleavage of aromatic ring bonds or of two or more C—C bonds to the same carbon atom may be forbidden. After the first stage, each structure–spectrum pair is then investigated in turn as the program seeks evidence in the spectrum for each fragmentation generated in ALLBREAKS, making allowance for the transfer of up to two hydrogen atoms either way. Finally, the evidence is collected and correlated. Common fragmentation modes in a class are grouped together and a summary output is provided.

There is significant ambiguity in the results obtained. Most ions in the spectrum will have alternative explanations, with the number of alternatives generally increasing as the size of the charged fragment decreases. Ambiguities can only be resolved by examination of the results for compounds possessing substituents on skeletal atoms which are lost in one process but not in an alternative process. This method has the potential drawback that a particular substituent may in reality be directing the fragmentation of the molecule. To try to avoid this possibility the INTSUM output is also examined manually for ambiguities. In the example given in Fig. 7.13 both the processes leading to 14 carbon fragments are still found to occur after manual removal of ambiguities.

The Ion Generator program developed at the University of Geneva, is designed to rationalize the formation of ions in a mass spectrometer on the basis of electron book-keeping theory. Electron book-keeping, a term coined by Djerassi (44), implies that discussions of reactions occurring in the MS should take proper account of the movement of electrons during these reactions. However, there are considerable differences between normal electron book-keeping theory and the version employed by the Ion Generator. In the conventional theory, mass spectroscopists use mechanisms to make rationalizations of ions observed in mass spectra. In the Ion Generator, mechanistic theory is used to make predictions of probable fragmentations which are then checked against those observed. Thus, the contribution made by the Ion Generator program is the specification of primitive mechanisms that can be applied again and again in ionic structures so long as the required motivating environment exists in the structure. Since new primitives can easily be introduced into the program when necessary, the program should gradually become a better representation of the knowledge of fragmentation processes.

In early work (28a, 28b) only four primitives were used. These were ionization with charge localization, bond homolysis β to a radical site, bond formation between two adjacent radical sites and transfer of a hydrogen atom to a radical site via cyclic transition states of various sizes. There were no primitives based upon heterolytic processes. The primitives listed describe the operations and, to a reasonable extent, the environments necessary for their motivation. Even with this limited number of primitives and with the further restriction that only primary ions, i.e. ions formed by direct fragmentation of the molecular ion, would be constructed, a considerable portion of the spectrum of a compound such as 5-diethylaminopentan-2-one could be explained (28a). A further feature of this program was the ability to

Process label	Observed/ total	Retention of n skeletal carbon atoms $n=$	Symbolic description	Ambiguity	Alternative explanations	%Σ range	Most frequent hydrogen transfers
20L	43/47	14		Yes	2L/11L	3·7–0	–1, –2
2L/11L	45/47	14		Yes	20L	6·4–0	0, –1, –2

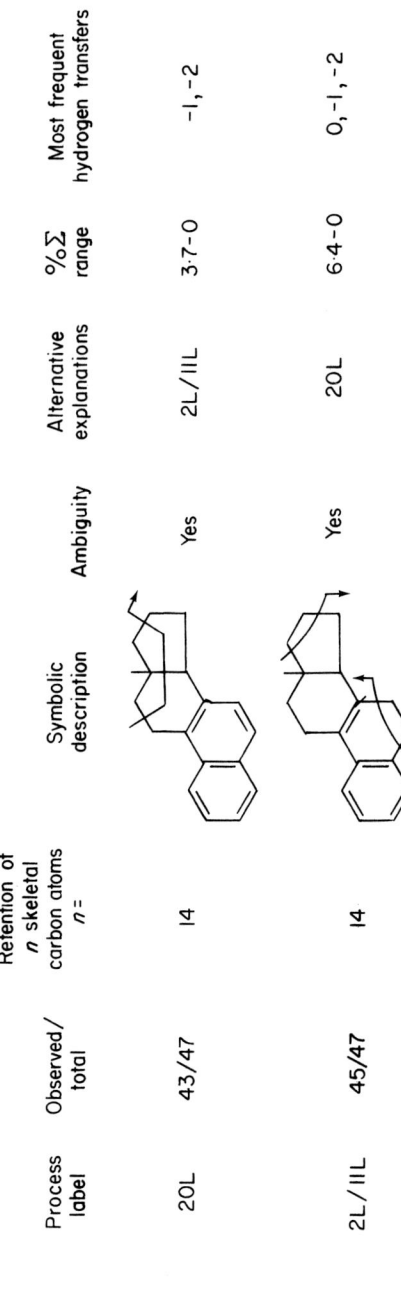

Fig. 7.13. INTSUM summary of two fragmentation processes shown by ≥85% (≥40/47) of the simple estrogens. Reprinted with permission from ref. 29. Copyright by the American Chemical Society.

specify an ionization threshold value so that, in effect, the particular localized electron which was to be removed to form a molecular ion could be specified. For 5-diethylaminopentan-2-one the ion generator proposed two such molecular ion structures, from which it found 37 primary ions; 22 of these had the charge on the nitrogen atom and the other 15 had the charge on the oxygen atom.

When the program was later extended to include the fragmentation of daughter ions and even second generation ions as well as the molecular ion, it was found necessary to take a much stricter look at the motivating structural environments, particularly if too many mechanisms were not to be generated. The practical realization of this new strategy was a new set of 16 primitives which are, in effect, concerted sets of four basic reactions, viz. homolysis, heterolysis, rearrangement and bond formation (28d, 28e) (see

A—B⌢C	⟶	A—Ḃ	+Ċ	Homolysis
A⌢B—C	⟶	A=B	+C	Heterolysis
Ȧ⌢(H)B	⟶	HA—Ḃ		Rearrangement
Ȧ—Ḃ	⟶	A=B		Bond Formation

Fig 7.14. Four basic reactions used in new strategy for Ion Generator (From ref. 28e.)

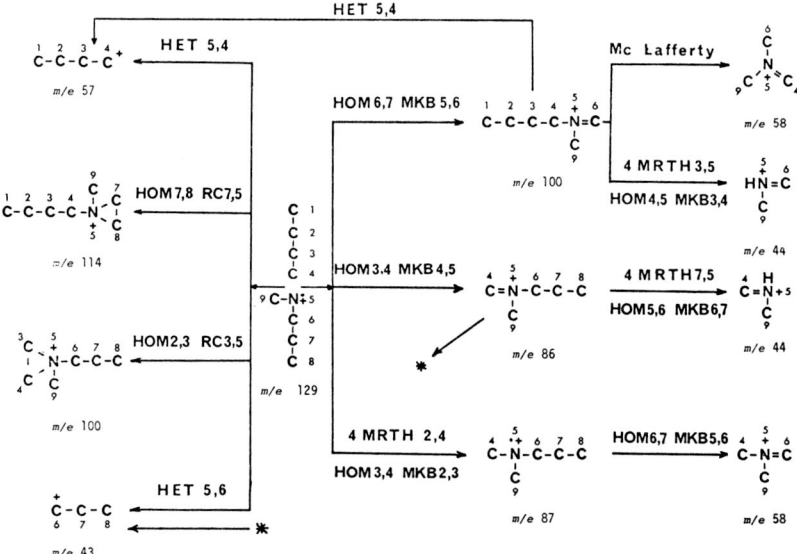

Fig. 7.15. Complete fragmentation pattern for methyl-n-propyl-n-butylamine proposed by Ion Generator program (compare spectrum in Fig. 7.16.) HET = heterolysis; HOM = homolysis; RC = ring closure; MKB = bond formation; 4MRTH = hydrogen transfer via 4-membered transition state. (From ref. 28d.)

Fig. 7.14). A simple example of one of these primitives is α-cleavage which, in fact, comprises the basic reactions of homolysis and bond formation (eqn. 7.2):

$$R' - \dot{A} - B - C - R'' \rightarrow R' - A = B + \dot{C} - R'' \quad (7.2)$$

Results obtained using this new set of primitives are summarized in Table 7.5 and the complete fragmentation pattern proposed for methyl-n-propyl-n-butylamine is shown in Fig. 7.15.

TABLE 7.6

Ion Generator results obtained using new strategy [a]

Compound	Number of different mechanisms proposed for each compound	% ion current explained by proposed mechanisms
Methyl-n-propyl-n-butylamine	14	80
Ethyl sec-butyl ether	15	85
Octan-3-ol	83	91
Dodecan-6-one	139	88

[a] (From ref. 28e.)

7.5.5 Summary

Current DENDRAL applications involve the use of INTSUM to analyse the mass spectra of a number of steroid classes as well as those of macrolide antibiotics, insect juvenile hormones and phytoecdysones. Again, the Planner program is being used to screen the mass spectra of marine sterols to identify quickly the spectra of known compounds and to suggest structures for new compounds. Unfortunately, many of the DENDRAL programs are large, requiring considerable core storage, and demand considerable computational effort. For example the application of the Planner program to the more difficult problems in estrogen structural analysis (37) required up to 1000 s of IBM 360/67 time. The majority of the DENDRAL programs are written in the LISP programming language, a high level language particularly suited to program development in this type of work but slow in execution. These programs are also consequently extremely difficult to transfer to other computer facilities.

However, the Dendral programs besides many others, are available to outside users via the TYMNET and ARPANET computer networks. The DENDRAL programs are mounted on the SUMEX—AIM (Stanford University Medical Experimental Computer—Artificial Intelligence in

214   COMPUTERS IN MASS SPECTROMETRY

Medicine) DEC KI-10 computer. The SUMEX facility was established to promote the sharing of such complex programs which would be difficult to mount on another computer system. Programs such as Congen, in particular, have already been used widely by a number of investigators. Many more details and a frank assessment of some of the problems of a networking system such as SUMEX—AIM are described in a paper by Carhart *et al.* (45).

7.6 USE OF METASTABLE DATA IN INTERPRETATIVE SCHEMES

A number of authors have recognized the value of metastable data in supplementing nominal and accurate mass data in interpretative schemes, particularly in mixture analysis (16, 38, 46). By defining the routes by which ions decompose, information on the manner in which the fragments represented by these ions are joined together in the complete molecule may be obtained. Practical interpretative schemes incorporating metastable data are rare, however, mainly because of the extended experimentation required to obtain metastable ions and the low sensitivity for metastable ions as compared with normal ions. It may be, as has been proposed in the DENDRAL system,

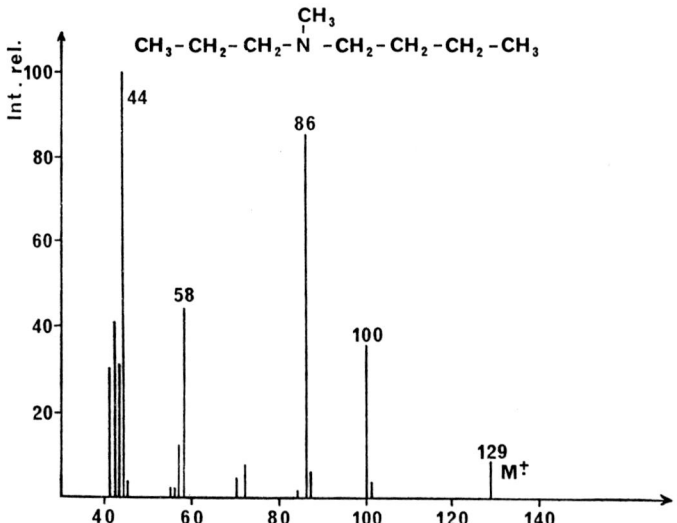

Fig. 7.16. Low resolution 70 ev spectrum of methyl-n-propyl-n-butylamine. (From ref. 28d.)

that the most practical use of metastable data will be one in which a preliminary examination of normal data suggests a limited number of observations of metastable data needed to elucidate the structure.

The simplest application of a computer to the interpretation of metastable data is in the form of a program to correlate the mass of an observed metastable ion m^* with the appropriate transition between ions observed in the normal spectrum according to eqn. 7.3 (47, 48):

$$m^* = m_2^2/m_1 \quad (\text{transition } m_1^+ \rightarrow m_2^+ + m_3). \tag{7.3}$$

A more complex approach to the interpretation of metastable data, applied to a high resolution scan recorded on a photoplate, was described by Mancuso et al. (49). Using eqn. 7.3 this program calculates all possible combinations of m_2 and m_1 for the metastable peak m^* using only masses for m_2 and m_1 that are present in the spectrum. The values of m_1 and m_2 chosen must meet a number of further conditions, viz; (a) the product of the normal peak intensities at m_1 and m_2 must exceed a certain minimum value; (b) the composition of the daughter ion (m_2) must be contained within the composition of the parent ion (m_1); and (c) the elimination of two radicals (as opposed to two neutral molecules) in consecutive steps is not allowed. Thus, only daughter ions of odd nominal mass can be considered to have arisen from nitrogen-free ions of odd nominal mass. For nitrogen-containing ions, the same principle is followed, allowing for the fact that an odd number of nitrogens reverses this rule. On this basis, the program prints the most likely assignments for each metastable peak observed.

As mentioned in Chapter 3, use of a defocusing technique to observe metastable transitions allows unambiguous assignment of the transitions as well as affording an increase in sensitivity. This approach was adopted by Barber et al. (46). The information used by their metastable analysis program comprised a list of all transitions detected together with the accurate mass of each ion in the spectrum. The transitions were arranged as fragmentation pathways from the molecular ion from which redundant pathways were then removed. In removing redundant pathways, longer pathways were, in general, considered as being more reliable since some consecutive processes appearing as a single loss, e.g. the consecutive loss of CH_3 and H_2O appearing as a direct loss of 33 a.m.u., has been observed. Finally, the possible atomic compositions for the units lost were calculated from the accurate mass data and then printed out together with that of the final daughter ion. Figure 7.17 shows the print-out for cortisone with the fragmentation pathways listed as atomic composition units lost from the molecular ion. Defocusing techniques were also used to acquire metastable data for the direct analysis of unseparated mixtures of extrogens (p. 204).

Fragmentation pathways from molecular ion are as follows

```
360 | 256 |
360 | 257 | 149 | 107 | 79 | 77 |
360 | 301 |
360 | 300 | 122 | 107 | 79 | 77 |
360 | 330 | 122 | 107 | 79 | 77 |
360 | 342 | 256 |
360 | 258 | 189 | 173 | 145 |
360 | 257 | 159 |
360 | 257 | 173 | 145 |
```

Remaining pathways after partial removal of redundancies are given below

```
360 | 301 |
360 | 330 | 122 | 107 | 79 | 77 |
360 | 330 | 256 |
360 | 330 | 257 | 149 | 107 | 79 | 77 |

| C 2 H 3 O 2 | C 19 H 25 O 3


| C | H 2 O |  | C 12 H 16 O 3 | C | H 3 |
| C | O |  | C O H 2 | C 6 H 5


| C | H 2 O |  | C 3 H 6 O 2 | C 17 H 20 O 2


| C | H 2 O |  | C 3 H 5 O 2 | C 7 H 8 O |
| C 3 H 6 | C | O |  | C 8 H 2 | C 6 H 5
```

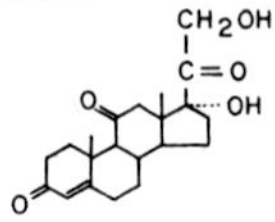

Fig. 7.17. Fragmentation pathways for cortisone elucidated from metastable data. (From ref. 46.)

7.7 SUMMARY

After the original work in the mid-1960s, work on interpretative schemes, particularly those making use of accurate mass data, lost a lot of its initial impetus. This can be attributed to a number of factors. The complexity of the formal programs when applied to all but the simplest cases, the expense and expertise needed to obtain complete high resolution spectra and the relative insensitivity of high resolution operation. In addition, other techniques such as library searching of nominal mass spectra appeared to produce useful results much more readily.

The formalization of knowledge relating to the fragmentation of compounds in the MS is a very fundamental and correspondingly difficult task. However, more recently, advances have been seen in this field through the use of the Ion Generator program of Buchs and Delfino (28) and INTSUM program of the Stanford school (29). Both these programs have been discussed in the section describing the DENDRAL project. Again, despite the experimental difficulties involved, it would be wrong to ignore the extra precision that accurate mass measurement data can give to the structural investigation of unknowns whether simple or complex. It is to be hoped that recently introduced techniques for obtaining accurate mass data at low resolving power and therefore at high sensitivity will be of value in this field (30).

REFERENCES

1. Pettersson B. and Ryhage R. (1967). *Analyt. Chem.* **39**, 790.
2. Pettersson B. and Ryhage R. (1966). *Ark. Kemi* **26**, 293.
3. Crawford L. R. and Morrison J. D. (1971). *Analyt. Chem.* **43**, 1790.
4. Crawford L. R. and Morrison J. D. (1969). *Analyt. Chem.* **41**, 995.
5. O'Brien J. F. and Morrison J. D. (1973). *Aust. J. Chem.* **26**, 785.
6. Jardine A., Reed R. I. and Silva M. E. S. F. (1973). *Org. Mass Spectrom.* **7**, 601.
7. Reimendal R. and Sjövall J. B. (1973). *Analyt. Chem.* **45**, 1083.
8. Gray N. A. B. and Grönneberg T. O. (1975). *Analyt. Chem.* **47**, 419.
9. Gray N. A. B., Zoro J. A., Grönneberg T. O., Gaskell S. J., Cardoso J. N. and Eglinton G. (1975). *Analyt. Letts* **8**, 461.
10. Gray N. A. B. (1975). *Org. Mass Spectrom.* **10**, 507.
11. McLafferty F. W. (1963). "Advances in Chemistry", Series No. 40. American Chemical Society, Washington.
12. Biemann K., Cone C., Webster B. R. and Arsenault G. P. (1966). *J. Am. chem. Soc.* **88**, 5598.
13. Senn M., Venkataraghavan R. and McLafferty F. W. (1966). *J. Am. chem. Soc.* **88**, 5593.
14. Barber M., Powers P., Wallington M. J. and Wolstenholme W. A. (1966). *Nature* **212**, 784.
15. Van't Klooster H. A., Vaarkamp-Lijnse J. S. and Dijkstra, G. (1974). *In* "Advances in Mass Spectrometry". (Ed. West A. R.), Vol. 6, p. 1027. Applied Science Publishers, Barking.
16. Wipf H-K., Irving P., McCamish M., Venkataraghavan R. and McLafferty F. W. (1973). *J. Am. chem. Soc.* **95**, 3369.
17. (a) Nau H., Kelley J. A. and Biemann K. (1973). *J. Am. chem. Soc.* **95**, 7162.
    (b) Kelley J. A., Nau H., Forster H-J. and Biemann K. (1975). *Biomed. Mass Spectrom.* **2**, 313.
    (c) Nau H., Forster H-J., Kelley J. A. and Biemann K. (1975). *Biomed. Mass Spectrom.* **2**, 326.
18. Nau H. and Biemann K. (1974). *Analyt. Chem.* **46**, 426.
19. Dayhoff M. O. and Eck R. V. (1970). *Comput. biol. Med.* **1**, 5.
20. Venkataraghavan R., McLafferty F. W. and Van Lear G. E. (1969). *Org. Mass Spectrom.* **2**, 1.
21. Mandelbaum A., Fennessey P. and Biemann K. (1967). 15th Annual Conference on Mass Spectrometry and Allied Topics, Denver. Paper 38, p. 111.

22. Koo D. H. K. and Sedgwick R. D. (1974). *In* "Advances in Mass Spectrometry" (Ed. West A. R.), Vol. 6, p. 1019. Applied Science Publishers, Barking.
23. Hilmer R. M. and Taylor J. W. (Unpublished data.)
24. Biemann K. and McMurray W. (1965). *Tetrahedron Letts* 647.
25. Adams R. A. and Sedgwick R. D. (1974). *Org. Mass Spectrom.* **9**, 884.
26. Biemann K. and Fennessey P. V. (1967). *Chimia* **21**, 226.
27. Biemann K., McMurray W. J. and Fennessey P. V. (1966). *Tetrahedron Lett* 3997.
28. (a) Delfino A. B. and Buchs A. (1972). *Helv. Chim. Acta* **55**, 2017.
    (b) Delfino A. B. and Buchs A. (1974). *Org. Mass Spectrom.* **9**, 459.
    (c) Delfino A. B. and Buchs A. (1973). "Topics in Current Chemistry" **39**, 109. Springer Verlag, Berlin.
    (d) Delfino A. B., Buchs A. (unpublished data.)
    (e) Delfino A. B. (1974). "The Ion Generator, a Heuristic Program in Organic Mass Spectrometry", Ph. D. Thesis, University of Geneva, Geneva.
29. Smith D. H., Buchanan B. G., White W. C., Feigenbaum E. A., Lederberg J. and Djerassi C. (1973). *Tetrahedron* **29**, 3117.
30. Aspinal M. L., Chapman J. R., Compson K. R., Hazelby D. and Riddoch A. (1973). 21st Annual Conference on Mass Spectrometry and Allied Topics, San Francisco. Paper T13, p. 471.
31. Slagle J. R. (1971). "Artificial Intelligence, the Heuristic Programming Approach". McGraw Hill, New York.
32. Duffield A. M., Robertson A. V., Djerassi C., Buchanan B. G., Sutherland G. L., Feigenbaum E. A. and Lederberg J. (1969). *J. Am. chem. Soc.* **91**, 2977.
33. Lederberg J., Sutherland G. L., Buchanan B. G., Feigenbaum E. A., Robertson A. V., Duffield A. M. and Djerassi C. (1969). *J. Am. chem. Soc.* **91**, 2973.
34. Schroll G., Duffield A. M., Djerassi C., Buchanan B. G., Sutherland G. L., Feigenbaum E. A. and Lederberg J. (1969) *J. Am. chem. Soc.* **91**, 7440.
35. Buchs A., Duffield A. M., Schroll G., Djerassi C., Delfino A. B., Buchanan B. G., Sutherland G. L., Feigenbaum E. A. and Lederberg J. (1970). *J. Am. chem. Soc.* **92**, 6831.
36. Buchs A., Delfino A. B., Duffield A. M., Djerassi C., Buchanan B. G., Feigenbaum E. A. and Lederberg J. (1970). *Helv. Chim. Acta* **53**, 1394.
37. Smith D. H., Buchanan B. G., Engelmore R. S., Duffield A. M., Yeo A., Feigenbaum E. A., Lederberg J. and Djerassi C. (1972). *J. Am. chem. Soc.* **94**, 5962.
38. Smith D. H., Buchanan B. G., Engelmore R. S., Adlercreutz H. and Djerassi C. (1973). *J. Am. chem. Soc.* **95**, 6078.
39. Meyerson S. (1955). *Appl. Spectrosc.* **9**, 120.
40. Dromey R. G., Buchanan B. G., Smith D. H., Lederberg J. and Djerassi C. (1975). *J. org. Chem.* **40**, 770.
41. Buchs A., Delfino A. B., Djerassi C., Duffield A. M., Buchanan B. G., Feigenbaum, E. A., Lederberg J., Schroll G. and Sutherland G. L. (1971). *In* "Advances in Mass Spectrometry" (Ed. Quayle A.), Vol. 5, p. 314. Institute of Petroleum, London.
42. Masinter L. M., Sridharan N. S., Lederberg J. and Smith D. H. (1974). *J. Am. chem. Soc.* **96**, 7702.
43. Carhart R. E., Smith D. H., Brown H. and Djerassi C. (1975). *J. Am. chem. Soc.* **97**, 5755.
44. (a) Budzikiewicz H., Djerassi C. and Williams D. H. (1964). "Interpretation of Mass Spectra of Organic Compounds". Holden-Day, San Francisco.
    (b) Budzikiewicz H., Djerassi C. and Williams D. H. (1964). "Structure Elucida-

tion of Natural Products by Mass Spectrometry", Vols 1 and 2. Holden-Day, San Francisco.
(c) Budzikiewicz H., Djerassi C. and Williams D. H. (1967). "Mass Spectrometry of Organic Compounds". Holden-Day, San Francisco.
45. Carhart R. E., Johnson S. M., Smith D. H., Buchanan B. G., Dromey R. G. and Lederberg J. (August 1975). 170th Annual Meeting of the American Chemical Society, Chicago.
46. Barber M., Green B. N., Wolstenholme W. A. and Jennings K. R. (1968). *In* "Advances in Mass Spectrometry" (Ed. Kendrick E.), Vol. 4, p. 89. Institute of Petroleum, London.
47. Rhodes R. E., Barber M. and Anderson R. L. (1966). *Analyt. Chem.* **38**, 48.
48. Brady L. E. (1971). *J. chem. Educ.* **48**, 469.
49. Mancuso N. R., Tsunakawa S. and Biemann K. (1966). *Analyt. Chem.* **38**, 1779.
50. Smith D. H., Masinter L. M. and Sridharan N. S. (1974). *In* "Computer Presentation and Manipulation of Chemical Information" (Eds Wipke W. T., Heller S. R., Feldman R. T. and Hyde E.) p. 287. John Wiley and Sons, Chichester.

8
Quantitative Analysis

8.1 INTRODUCTION

One approach to quantitative analysis by mass spectrometry uses the heights of peaks taken from a conventional scan as a basis for measurements. This method, which generally requires no additional hardware, is perhaps the simplest approach experimentally. However, additional software to process the data to give quantitative information has been developed and this is discussed in Sections 8.2–8.4.

An alternative approach is to use the mass spectrometer as a more specific detector so that it monitors only those mass values of relevance to the determinations. This approach was developed as "selected ion monitoring" by Sweeley *et al.* (1) as a technique for the determination of stable isotope abundance in compounds eluted from a gas chromatograph (GC) and for qualitative and quantitative analysis at very high sensitivities. A large proportion of the application of mass spectrometry in the fields of biochemical, pharmacological and environmental research is based on this technique. Its implementation requires both new hardware and software and these are both discussed in Sections 8.5–8.9. A separate section, 8.10, considers the quantitative analysis of inorganic materials using spark source MSs of Mattauch–Herzog geometry.

8.2 QUANTITATIVE DATA FROM SCANS OF UNRESOLVED MIXTURES

In quantitative analysis from scan data, the experimental data may be represented by the following set of linear simultaneous eqns:

$$\begin{aligned}
i_1 &= a_{11}p_1 + a_{12}p_2 + \cdots + a_{1n}p_n \\
i_2 &= a_{21}p_1 + a_{22}p_2 + \cdots + a_{2n}p_n \\
&\vdots \\
i_m &= a_{m1}p_1 + a_{m2}p_2 + \cdots + a_{mn}p_n
\end{aligned} \quad (8.1)$$

where i_m is the measured peak height at mass m in the mixture spectrum, p_n is the partial pressure of component n and a_{mn} is the peak height at mass m due to unit pressure of component n.

The earliest and still a major application of this type of analysis is to hydrocarbon mixtures encountered in the oil industry. In this case, the requirement is usually for an analysis of component groups, e.g. weight percentage of paraffins, cycloparaffins and benzenes, where each group can be characterized by the intensity sum at selected masses, e.g. m/e 71 + 85 + 99 + 113 for paraffins. Although both low resolution (2–4) and high resolution methods (5–6) have been presented for this type of mixture analysis, there is an increasing tendency towards use of the latter because of their higher specificity.

The use of one characteristic mass or mass series for each component means that there are now only n equations for n unknowns. Thus, eqns 8.1 becomes eqns 8.2 where the i and a values now refer to each mass series:

$$\begin{aligned} i_1 &= a_{11}p_1 + a_{12}p_2 + \cdots + a_{1n}p_n \\ i_2 &= a_{21}p_1 + a_{22}p_2 + \cdots + a_{2n}p_n \\ &\vdots \\ i_n &= a_{n1}p_1 + a_{n2}p_2 + \cdots + a_{nn}p_n. \end{aligned} \qquad (8.2)$$

In the notation of matrix algebra, these equations may be written as:

$$I = AP \qquad (8.3)$$

and the solution as:

$$P = A^{-1}I \qquad (8.4)$$

where A^{-1} is the inverse matrix. The inverse matrix is developed by calibration with pure compounds (7) or mixtures of known composition (8).

In other cases, where, because of fragmentation pattern overlap, the analysis required does not permit the direct choice of suitable mass values, all the data is used to solve eqns 8.1. Since there are in general, more equations than unknowns ($m > n$), some function of the difference between calculated and observed values of i must be minimized. This is generally the least squares criterion (9) where the quantity given by eqn. 8.5 is to be minimized. Examples of the use of this method have been given by Raimondi et al. (10) in residual gas analysis and by Barnard (7). The discussion of hydrocarbon analysis given by Barnard (7) contains an excellent description of the methods of calculation used in quantitative analysis from scan data:

$$\sum_{r=1}^{m} (a_{r1}p_1 + a_{r2}p_2 + \cdots + a_{rn}p_n - i_n)^2. \qquad (8.5)$$

Effective use of the methods described so far requires considerable previous knowledge of the qualitative composition of the sample, as the presence of unexpected components or background causing interference may lead to incorrect results.

As previously mentioned, one way of minimizing this interference is by the use of data recorded at high resolution. An alternative approach is the stepwise regression method described by Tunnicliff and Wadsworth (12). This program is based on a mathematical method of choosing from among a large library of spectra such that the sum of the individual spectra after multiplying by the proper concentration factors gives the best least squares fit to the sample spectrum. In this method, an additional library spectrum is selected at each step to give the greatest improvement in the variance of the fit of the predicted spectrum to the sample spectrum (Table 8.1). The process of selecting the best set of spectra to describe the sample spectrum may be improved by preferentially entering any known or suspected components and then allowing the regression procedure to find any unknown components. There is no theoretical limit to the number of compounds that may be considered by this procedure, although the number of compounds found cannot exceed the number of masses. Thus, the library described by Tunnicliff and Wadsworth may contain as many as 150 reference spectra with data for 110 masses for each spectrum.

TABLE 8.1

Operation of stepwise regression procedure of Tunnicliff and Wadsworth[a]

Compound	Actual %	Found %				
		Step 1	Step 2	Step 3	Step 4	Step 5
n-Butane	40.0	64.5	62.8	53.5	40.2	40.1
Isobutane	15.0				14.1	14.8
Nitrogen	15.0					15.1
Carbon monoxide	20.0		23.6	22.0	23.0	20.1
Acetone	10.0			13.0	12.2	9.8
Variance, %		33.9	24.6	17.6	11.2	1.0

[a] (From ref. 12.)

The use of scan data in petroleum analysis where the number of components of the mixture exceeds the number of analytical measurements has also been described by Tunnicliff and Wadsworth (13). In this case, the spectra of standard hydrocarbons are used to train a pattern recognition procedure, using a least squares method to determine the appropriate weight values.

This procedure and results obtained from it are fully described in Chapter 6 (p.167).

8.3 QUANTITATIVE DATA FROM REPETITIVE SCANS OF SEPARATED MIXTURES

Data from repetitive scans taken on mixtures separated by gas chromatography has been used for quantitative analysis, for example, in the field of steroid analysis by Sjövall and coworkers (14) and in the analysis of acidic urinary metabolites by Sweeley and coworkers (35). Such scans have either been scans of the entire mass range or scans over a limited mass range for increased sensitivity. Sjövall describes two methods by which repetitive scan data may be used, quantitative data in each case being calculated from GC peak areas under mass chromatograms.

In the first method, the user inspects mass chromatograms of appropriate m/e values and selects sections for peak area determination and sections to be used as background. Peak areas are determined by the addition of intensity readings at appropriate m/e values over the GC peak. The mean intensity value for each mass in the background spectra is multiplied by the number of readings over the GC peak and this value is then subtracted. Peak area is converted to quantity by comparison with the area given by known amounts of the standard compound (external standard). An internal standard may also be monitored by selecting an appropriate m/e value, and data from this peak can be used to correct for differences in injected volumes, etc. A somewhat similar system has been described by Baczynskyj et al (15) where narrow range magnet scans are initiated and recorded only when the sample and internal standard peaks are being eluted. These authors have used the method to measure prostaglandins in biological fluids using a deuterated analogue as internal standard.

The second method described by Sjövall is considerably more automated and operates with internal or external standards or both. The information required by the program is: (a) the analysis numbers of the external standard and of the samples; (b) amount of external standard; (c) names of compounds to be determined and relative retention time ranges within which each compound might be expected; and (d) up to 10 typical m/e values for each compound to be determined. The computer locates the peaks in individual chromatograms, notes retention times and calculates peak areas. A baseline which may be horizontal or may have a constant slope, is constructed and background is subtracted. The peak area in individual mass chromatograms is converted to the equivalent total ion current (TIC) peak area. This can be done from a knowledge of the contribution of each fragment ion to the total ionization of a compound, previously determined by the computer from an averaged mass spectrum of the same compound. This conversion permits a comparison between peak areas of different compounds on the basis that equal amounts of structurally similar compounds will give approximately equal TICs. By this means, quantitative data, although only approximate, may be obtained even when there is insufficient pure compound for calibration.

A similar method, but using accurate mass data, has been applied by Schuetzle *et al.* (16) to the direct analysis or both organic and inorganic air pollutants. In this method, samples collected from filters are introduced directly into the MS using a temperature programmed solid probe and high resolution spectra are acquired continuously during the volatilization of the sample into the source. Computer identification of particular components from this recorded data can be made on the basis of several tests:

(a) The presence of a major identifying mass ($M1$) (Fig. 8.1).
(b) The temperature at which the component was detected (Fig. 8.1).
(c) The presence of one or two other major identifying masses ($M2$ and $M3$) at this temperature.
(d) The possible existence of the component as an aerosol as determined by vapour pressure calculations.

Comparisons of unknown masses with known masses from an air particulate mass file are made within limits of 0·003 a.m.u.

The integrated ion current of the major identifying mass (UNK) is then calculated and compared to the total integrated ion current of an internal

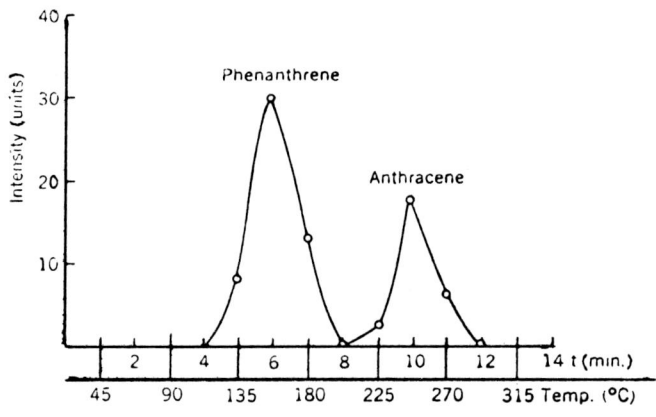

Mass thermogram for m/e: 178.078.

Fig. 8.1. (Reprinted with permission from ref. 16. Copyright by American Chemical Society.)

standard (STD). The quantity of material present is given by the formula:

$$\text{CONC}(\mu g/m^3) = (UNK)(X)/(STD)(RF)(AL)(M^3) \tag{8.6}$$

where X is the quantity of standard added to the sample in µg, RF is the sensitivity ratio expressed in terms of TIC for $M1$ compared to that of the internal standard, AL is the aliquot of sample used for analysis and M^3 is the volume of air sampled in cubic metres.

Another possible approach to quantitative data from repetitive low resolution scans is contained in the vector analogy method of Rosenthal and Bursey (17), already described as a method of sensitivity enhancement in Chapter 4 (p. 91).

8.4 STABLE ISOTOPE DETERMINATION FROM SCAN DATA

Analysis by mass spectrometry permits the use of stable isotopes rather than radioactive isotopes in metabolic studies. Thus, molecules differing in the number of heavy atoms incorporated can readily be distinguished and sometimes the position of a heavy atom may be directly determined from the isotope content of appropriate fragment ions. Sjövall and coworkers (14a) have written a program that calculates isotopic compositions by comparing the averaged mass spectrum (or partial mass spectrum) obtained by repetitive scanning over the peak of the labelled compound eluted from a GC with the averaged spectrum of the unlabelled standard compound. In addition to a background correction and corrections for natural abundance of stable isotopes, the program contains an optional correction for the presence of a P-1 peak.

With compounds containing from 0–15 heavy atoms, and when repetitive scanning of entire mass spectra is used, sample sizes of 0·1–1 µg are required for measurements on major fragment ions. A 100-fold increase of sensitivity can be obtained by repetitive accelerating voltage scanning over a limited mass range. An example of the output of this program is shown in Fig. 8.2.

Figure 8.2 also demonstrates the partial separation of differently labelled molecular species that occurs in gas chromatography. Because of this separation, no single scan can be representative of the overall isotopic distribution, hence the need for rapid repetitive scans covering the GC peak in this type of analysis. Dowie et al. (18) have investigated the errors involved in calculating isotopic distributions by averaging data from repetitive scans and the dependence of these errors on the number of scans averaged.

A high resolution scanning technique for ^{15}N measurement has been reported by Haddon and coworkers (19). This technique derives the isotope ratio from the abundance of the component peaks of a multiplet at a single nominal mass containing the required isotopes. The isotopic multiplet is scanned repetitively by varying the electric sector voltage of a double focusing MS and the recorded data is time averaged to obtain an enhanced signal to noise ratio and to correct for changing sample concentration.

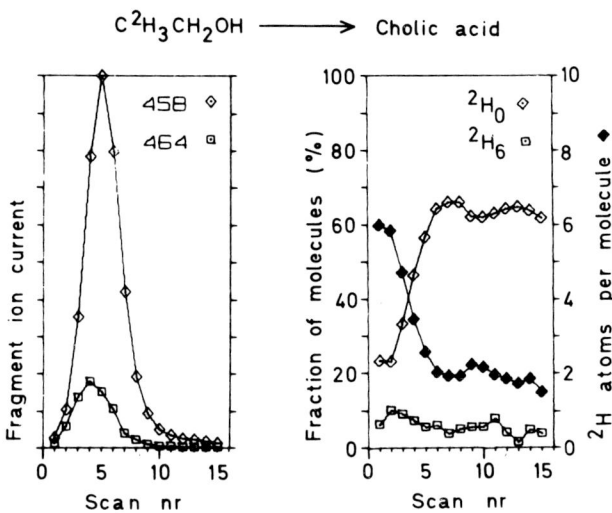

Fig. 8.2. Left panel: TIC chromatograms of m/e 458 and 464 representing M—180 of unlabelled and hexadeuterated cholic acid molecules formed during metabolism of $(2,2,2-{}^2H_3)$ ethanol in a bile fistula rat. Right panel: Isotope composition calculated by ISOTN from each individual mass spectrum taken in the analysis shown in the left panel. Plots of the percentage of unlabelled and hexadeuterated molecules and of the total deuterium excess at different parts of the GLC peak were obtained with program DIAGR. (From ref. 14a.)

As an example, Fig. 8.3 shows the result of time averaging 64 scans taken at a resolving power of 25 000 of the M + 1 peak of unlabelled creatinine ($C_4H_7ON_3$) introduced on the solids probe. The principal peaks at this mass are those due to natural ^{13}C and ^{15}N isotopes, so that, for a ^{15}N labelled sample, the ^{15}N level can be measured by considering the ^{13}C peak as an internal standard. The additional peak on the high mass side of the multiplet is dependent on the sample pressure and thus arises from an ion–molecule reaction. An advantage of the high resolution method over low resolution recording is that any contribution from such peaks is automatically removed. At low resolution, calibration with an unlabelled compound would remove this source of error only if the evaporation profile could be duplicated exactly for each sample.

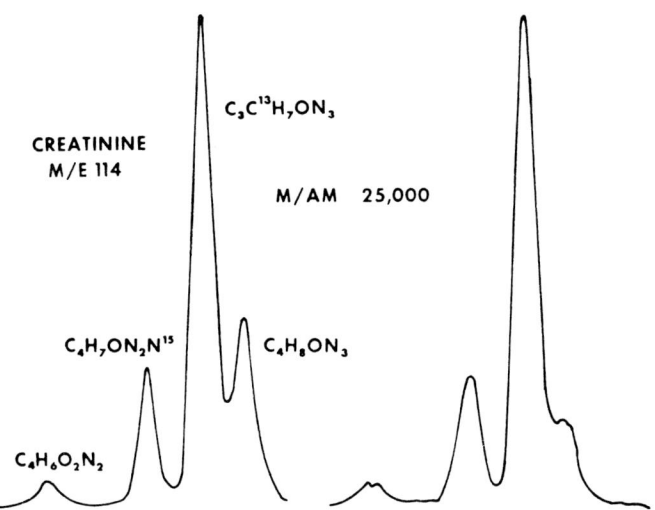

Fig. 8.3. Creatinine isotopic multiplet mass 114. Left scan: ion source pressure 2×10^{-5} Torr. Right scan: ion source pressure 3×10^{-6} Torr. Resolution: 25 000, 64 time-averaged scans. (Reprinted with permission from ref. 19. Copyright by American Chemical Society.)

Haddon et al. describe two methods of treating the time averaged scan data prior to isotope ratio calculation. The more complex method uses published procedures to fit Gaussian profiles to the individual peaks which may frequently be incompletely separated. A simpler method, used to calculate the amount of $^{15}N_2$ incorporated into creatinine, locates the valley between the $^{15}N_2$ component and the component used as internal standard and integrates each peak from a threshold of 5% of the peak maximum to the valley minimum. A correction for overlap assuming Gaussian profiles is applied and digital smoothing used to improve the quality of the raw data. Table 8.2 lists results for measurements at different levels of $^{15}N_2$ incorporation in the creatinine samples.

The method of Haddon affords routine measurements of excellent accuracy and precision on microgram samples. For example, the results of Table 8.2 indicate a detection limit of about 0·002 atom % excess for $^{15}N_2$ labelled creatinine. ^{15}N from labelling experiments may be measured using natural ^{13}C as the internal standard or ^{13}C may be measured in nitrogen containing compounds using natural ^{15}N as the internal standard. However, this method is inapplicable to compounds not containing heteroatoms. Given sufficient instrumental sensitivity, other workers (20) have demonstrated that high resolution scans covering a much wider mass range may be recorded using a data system to give stable isotope levels directly.

TABLE 8.2

Atom per cent excess $^{15}N_2$ in labelled creatinine isolated from human urine [a]

Sample	Atom % excess of $^{15}N_2$	Sm[b]
2–24	0·638	0·0037, 0·58 %
2–27	0·559	0·0019, 0·34 %
2–32	0·524	0·0031, 0·59 %
2–36	0·517	0·0044, 0·85 %
2–42	0·479	0·0034, 0·71 %
2–47	0·442	0·0029, 0·65 %
2–63	0·357	0·0025, 0·70 %

[a] (Reprinted with permission from ref. 19. Copyright by American Chemical Society.)
[b] Relative standard error, based on 8–10 separate 64-scan averages on the same sample. Total sample consumption of less than 25 μg.

8.5 SELECTED ION MONITORING

The MS may be used as a more specific detector without scanning so that, in the simplest case, only one mass value is continuously focused at the collector. Thus, a MS tuned so as to respond only at *m/e* 298 provides a reasonably specific detector for methyl stearate (parent ion mass = 298·2872). Operation at higher resolving power with more precise mass selection, e.g. at *m/e* 298·29, can give yet more specific detection, although with lower sensitivity. More commonly, perhaps four masses are monitored in a cyclic manner, with rapid switching from one mass to the next and a separate trace as output from each mass. Multiple ion monitoring, carried out in this way, can offer either further specificity by monitoring more than one mass for each compound or the ability to monitor more than one compound in each run (e.g. Fig. 8.6). However, because of the switching speeds involved, multiple ion monitoring with sector instruments has usually been effected by switching the accelerating voltage, although a more recent report describes magnet switching as well (21). Quadrupole mass filters are generally easier to operate in the multiple peak monitoring (MPM) mode than magnetic sector instruments because of the ease with which the rod voltages can be switched to transmit different masses.

The recording of only selected mass values also offers very high sensitivity compared with scanning because long time constants may be used. For example, in a 1 s/decade exponential magnetic scan at 1000 resolving power, the peak width is 0·434 ms. However, a MPM system, examining four masses during 1 s will sample each mass for at least 200 ms, depending on the time allowed for voltage settling. This potential increase in sensitivity is usually realized in data system operation by integration of the analogue output or by ion counting techniques. If a large number of masses are to be monitored, for example during a GC analysis, it is better, where possible, to monitor smaller subsets of these masses for the appropriate section of the run. In this way, the sampling period for each mass does not become too short. In addition to their use in the sensitive and selective detection of organic materials in complex environments, peak switching techniques have found application in the determination of both stable (22, 23) and radioactive isotopes (24–26).

8.6 CONTROL OF SECTOR INSTRUMENTS FOR SELECTED ION MONITORING

In peak switching with a magnetic sector instrument, the magnet current is set to monitor the lowest mass required using the normal accelerating voltage. The other masses required are then monitored by switching the accelerating voltage, bearing in mind that the mass of an ion reaching the collector is inversely proportional to the accelerating voltage which is used (see eqn. 8.7). Prior to the application of data systems, peak switching was effected using a hardware unit (accelerating voltage alternator, AVA) containing relays to switch between the normal accelerating voltage and smaller values obtained using a potential divider:

$$m/e = H^2 r^2 / 2V \quad \text{i.e.} \quad m_1 V_1 = m_2 V_2 \quad \text{(for constant magnetic field } H\text{)}.$$
(8.7)

The use of a data system to control the switching of a magnetic sector instrument offers a number of advantages over hardware control. Thus, the number of channels monitored, the switching rate and the sampling time may be modified without any hardware changes. The accelerating voltages required for the masses specified may be calculated and set up entirely under data system control without any operator intervention. Finally, real time correction for magnet drift may be effected using a data system. Figure 8.4 shows a typical system used with a sector instrument. Changes in the accelerating voltage are effected by means of a digital to analogue converter (D/A converter) which is used to control either all or part of this voltage (21, 23, 27, 28, 36). For example, the system described by Holmes and co-workers (27) uses a 15 bit D/A converter to control a ± 16.383 V range in 1 mV steps. This signal is amplified to cover a 1000 V range and is then

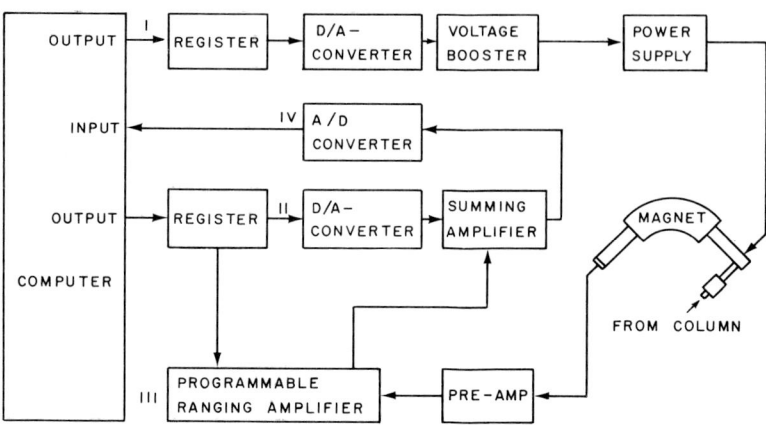

Fig. 8.4. Schematic diagram for computer-controlled selected ion monitoring and digital signal processing. I: Accelerating voltage control. II: d.c. offset for signal baseline control. III and IV: Analogue signal processing (see p. 21) (From ref. 36.)

added to a fixed 3000 V supply. Thus, the monitored mass can be varied over a range of approx. 30%, subdivided to a mass resolution of 1 part in 100 000.

A reasonably rapid settling time for the voltage after switching, is essential if too much time which would otherwise be used for sampling is not to be wasted in this way. Holmes states that a full 1000 V change had settled within 12 ms. Thus, his system allows 16 ms for the voltage to settle after switching before any sampling of the ion beam commences. Other published systems quote settling times of 2 ms (28) and approx. 10 ms (21) for single steps of 900 V and 400 V respectively. It may be noted that double focusing MSs have an advantage in this respect (Section 1.2.2) as it is only the much smaller electric sector voltage that must settle completely before sampling commences.

Perhaps the most important feature of any single or MPM system is the mass scale stability. It is important that if the system is tuned to monitor selected mass values at the beginning of a GC run it should still be tuned to these mass values when the components of interest elute several minutes later. It is also desirable that the system should not need re-tuning for subsequent runs using the same mass values. Drift in magnetic sector machines is usually associated with the magnet control circuitry because of the relatively large amounts of power dissipated, although present generation instruments generally offer excellent stability in this respect, and any drift of accelerating voltage supplies is comparatively very small.

In instruments where magnet drift has been found to be a serious problem, various forms of feedback control using the data system have been used to compensate for this. Holmes *et al* (27) monitored one mass value, preferably from an added calibration compound, by means of a stepped voltage sweep rather than at a single voltage value. From these samples taken across the peak, the peak position could be calculated, checked for drift and the D/A converter values for each ion adjusted accordingly. The sweep comprised 80 voltage increments generated at 2 ms intervals and covering a voltage range of ± 10 V and the subsequent calculations and voltage adjustments then took some 30 ms to complete. The correction procedure was tried in a test program and appeared to work. However, as the authors point out, the peak used for the corrections should be large enough so that statistical fluctuations in its position (p. 57) are considerably less than any drift it is hoped to compensate for.

A number of authors (21, 28, 29) have described feedback systems based on a Hall probe monitoring the magnetic field. In the system described by Hammar (28) the Hall voltage is measured on setting up the correct magnet current. If the magnetic field drifts during a run, the computer monitoring the Hall voltage will detect this drift and compensate by immediately calculating and setting new accelerating voltages. In the system used by Klein (29) for stable isotope ratio measurement, a magnet stabilization circuit based on feedback from a Hall probe is able to hold the magnetic field constant to within 10 p.p.m.

A voltage sweep controlled by the D/A converter for one or more of the masses monitored can also serve a number of other purposes (21, 27, 28, 30). For example, precise focusing of a mass or masses may be achieved by using the sweep to locate the centre of the peak and subsequently resetting the accelerating voltage to this value (21, 27, 28). This procedure is also of importance in setting up to monitor masses whose nominal mass only is known.

Another application of the voltage sweep is in peak monitoring at high resolution. Under high resolution conditions much greater specificity may be obtained. However, the reduced peak width means that any drift in peak position causes a relatively larger reduction in recorded peak intensity. Thus, it is often better to use a system that sweeps the mass peak and integrates the output from this sweep. Although some sensitivity is lost in this mode compared with integration at the peak top, the method has the outstanding advantage of being relatively immune to drift. A voltage sweep has also been used for to permit accurate mass measurement by peak switching (28) (p. 65) and to provide a visual display for use in setting up the instrument (27).

The question of mass range available has been briefly mentioned previously. Most installations have provided a mass range of approx. 30% but, by controlling the whole accelerating voltage, a much larger mass range is theoretically possible. The mass range required is one of the factors that will determine the resolution required from the D/A converter. At the highest resolving powers, monitoring will involve stepping the accelerating voltage to sweep through the peak and, if we accept that at least 10 samples are needed to cover a peak, then the effective resolution of the D/A converter must be equal to or better than 1/10 of the peak width. For an n bit converter, the effective resolution at an operating accelerating voltage V is given by:

$$R^* = \frac{10^6 \cdot V_{var}}{(2^n - 1) \cdot V} \text{ p.p.m.}$$

where V_{var} is the total voltage controlled by the DAC, i.e. $V =$ Full accelerating voltage -- V_{var}. Thus, at a resolving power of 10 000 (100 p.p.m.), we must have:

$$\frac{10^6 \cdot V_{var}}{2^n \cdot V} \leqslant 10$$

so that a 15 bit D/A converter will just provide a 30% mass range under these conditions.

Unfortunately, the accelerating voltage of a sector instrument may be switched over only a limited range without a significant loss of sensitivity. A system described by Sweeley and coworkers (21) controls and switches both the accelerating voltage and the magnetic field to overcome this problem. Instrumentation for this system includes a 14 bit D/A converter controlling the accelerating voltage between 3300 and 3700 V and a 10 bit D/A converter giving approximately linear control of the magnetic field between m/e 10 and m/e 750 (see Fig. 8.5).

A calibration procedure using perfluorokerosene (PFK) as reference

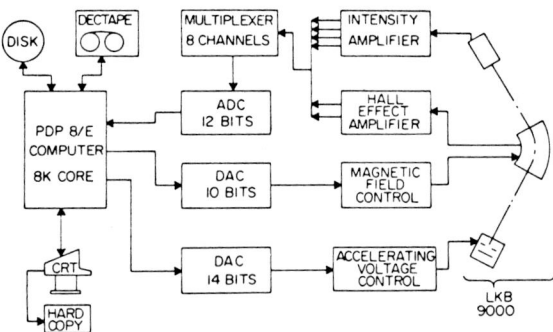

Fig. 8.5. Hardware configuration of GCMS data system for accelerating voltage and magnet field control. (Reprinted with permission from ref. 21. Copyright by American Chemical Society.)

compound first establishes a correlation between mass and the Hall probe voltage from a magnetic scan. Second, a correlation between D/A converter voltage and Hall voltage and therefore mass is established by ramping the magnetic field using the D/A converter. Both the scan and ramped sweep are upwards in mass. Masses to be monitored during an analysis are set in groups whose members lie within the acceptable range of accelerating voltage change, presently corresponding to a mass range of 11·4%. Switching between mass groups can be carried out automatically at preset times or manually by altering the magnet setting. Using the 10 bit D/A converter as control, the magnetic field is ramped to the desired mass, always approaching this mass from below and slowing the approach near to the desired setting. Failure to observe these last two points invalidates the calibration.

Setting the magnetic field in this way can be carried out to within 0·7 a.m.u. of the desired mass and the Hall voltage can be read and converted, using the calibration, to give a value of the actual mass setting to within 0·2 a.m.u. From a knowledge of this mass setting, the program calculates the accelerating voltage needed to focus each ion resulting in an approximate focusing at each mass value required. Fine focusing is automatically accomplished by a further routine. In this, the accelerating voltage is stepped in small increments across each ion peak, the ion beam sampled at each point and the centre of gravity of the peak calculated. The voltage is adjusted to this centre of gravity value and the ion beam sampled at this position. To avoid focusing on small peaks a minimum rate of change of intensity for successive data points must be detected before the fine focusing routine is activated, so that the initial part of any ion current trace representing a GC peak may be distorted until focusing is complete. For this reason, height rather than peak area is used as the quantitative measure. The program is capable of measuring one ion every 0·1 s with up to eight ions in each group. An error of up to 0·05 a.m.u. per switching cycle can be corrected by the fine focusing procedure.

An earlier system described by Sweeley and coworkers (30) used a data system to control and augment the performance of a commercial hardware accelerating voltage switching unit (AVA). The accelerating voltages required were manually set up using the hardware AVA to approximately correct values. To provide fine focusing, a signal with a maximum value of 10 V provided by a 10 bit D/A converter, was added to or subtracted from the main accelerating voltage as necessary under computer control. The routine used to set this additional voltage was the fine focusing routine described in the previous paragraph. A recent option in this fine focusing procedure allows independent focusing for each ion or the slaving of any ion from another ion. For example, a very weak peak may be focused entirely by reference to a much more intense one.

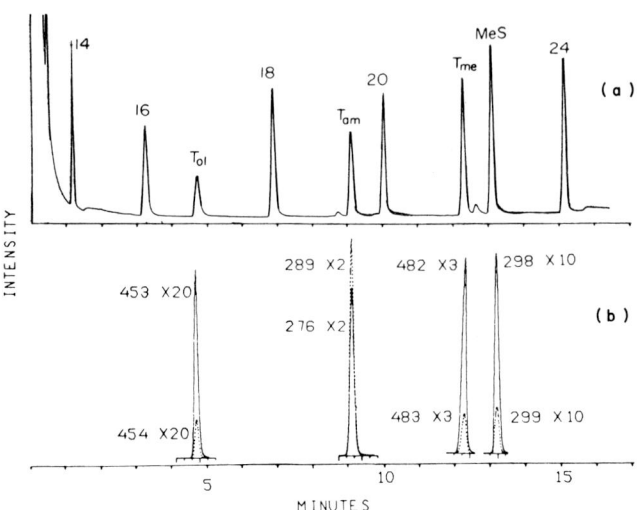
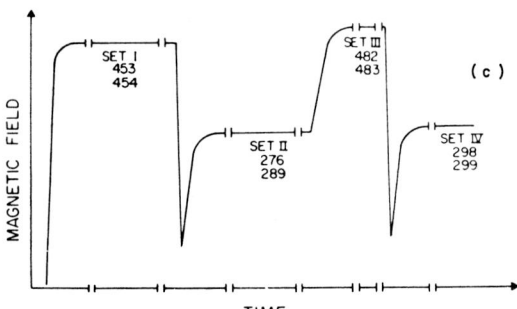

Fig. 8.6. (a) Gas–liquid chromatogram of a mixture of tryptophol (T_{ol}), tryptamine (T_{am}), 5-methoxytryptamine (T_{me}), methyl stearate (MeS), and hydrocarbons from C_{14} to C_{24} : 3% OV-17, 130–230°C at 4°/min. (b) Computer plots of ion intensities at m/e 453 and 454 (T_{ol}), 276 and 289 (T_{am}), 482 and 483 (T_{me}) and 298 and 299 (MeS), obtained in sequential selected ion monitoring analyses; gas–liquid chromatography conditions similar to those described above. (c) Simulated levels of magnetic field intensity for the sequential selected ion monitoring analyses. Numbers refer to m/e values focused in each set. (Reprinted with permission and adapted from ref. 21. Copyright by American Chemical Society.)

Data acquisition was also computer controlled in this AVA system, but peak switching rates were controlled using the hardware AVA. Another system described by Watson *et al.* (31) permitted data system control of the same commercial AVA unit and also used the data system to perform all data acquisition and data reduction processes. However, no automatic focusing facilities were available with this system.

8.7 CONTROL OF QUADRUPOLE INSTRUMENTS FOR SELECTED ION MONITORING

The use of a data system to control the switching of a quadrupole mass filter is relatively simple compared with the control of a magnetic sector instrument. The mass of the transmitted ion is proportional to the analyser voltages which are readily and rapidly switched. The analyser voltages are controlled using a D/A converter and, once a calibration has established a number of points on the mass–voltage relationship, any specified mass value may be set by a simple linear interpolation (eqn. 8.8):

$$V_c = V_1 + (M_c - M_1) \times (V_2 - V_1)/(M_2 - M_1). \tag{8.8}$$

The subscript c refers to the mass being set and 1 and 2 refer to reference masses bracketing this mass. Thus, a D/A converter with a resolution of 14 bits would be sufficient to provide a setting accuracy of 0·1 a.m.u. over the complete mass range for an instrument with unit resolution up to m/e 1000.

However, because of the very high switching rates possible with quadrupole instruments, each ion may be sampled over only a very few milliseconds with continuous cycling over the series of ions to be measured. Such a system can approach the ideal of simultaneous collection of ions available with double collector instruments, in which peak height ratio measurements are unaffected by instrumental instabilities or pressure changes (32). Again, excellent figures for mass scale stability (better than $\pm 0·1$ a.m.u. over eight hours) are now claimed for quadrupole analysers. With such high stabilities, any form of built-in correction for drift should be unnecessary. However, a stepped voltage sweep through the peak is still necessary for calibration and can be used for precise setting of the mass required. In addition, a number of authors have described quadrupole switching systems in which a voltage sweep has also been used as a basis for data acquisition from sample peaks (22, 23). These include an integrated circuit stable isotope ratiometer–multiple ion detector unit (SIRMID) able to control the switching of either quadrupole or sector instruments (22). Data acquisition procedures with this latter unit are described on page 244.

Perhaps the major advantage of quadrupole MSs in multiple ion monitoring is that they can be operated without severe sensitivity losses over very wide voltage ranges. Thus, any ions may be measured together in a switching cycle and need not lie within a given percentage mass range.

8.8 DATA ACQUISITION FROM SELECTED ION MONITORING EXPERIMENTS

In principle, data acquisition from single or multiple ion monitoring experiments can be carried out with the same standard analogue to digital converter (A/D converter) system already described in Section 2.2.6. However, there is one important aspect in which this data differs from that acquired by scanning. This is the large dynamic range of the intensity data.

For example, in the single ion monitoring mode, data may be displayed using an output filter which is just sufficiently fast to reproduce the GC peak shape, say 0·2 Hz ($\tau = 0\cdot 8$ s). Under these conditions and using an amplifier input resistor of 5×10^7 Ω, a typical amplifier noise level has been measured as 17 μV peak to peak. With a multiplier gain of 10^5, a 34 μV signal would correspond to a collector current of $6\cdot 8 \times 10^{-18}$ A or just over 40 ions/s. The percentage standard deviation (SD) of this signal will be given by: %SD = $100\,(2n\tau)^{-\frac{1}{2}}$, where n is the number of ions/s. Thus, in this case: SD = 12·5% so that we may expect, with 95% confidence, that the signal level will fall between the limits of $\pm 25\%$, i.e. 25·5–42·5 μV. Such a signal should be readily detectable above the background noise level and an A/D converter with a resolution of the order of 10 μV will be required. With a maximum signal level of 10 V, this corresponds to a dynamic range of 10^6 or 20 bits (cf. 12 or 14 bits in scanning p. 26).

A 34 μV signal recorded at a gain of 10^6 would only correspond to just over 4 ions/s (or nearly 7 ions in a period equal to 2τ s) and now the variation in the signal level will be much greater. Thus, the Poisson distribution (see Appendix) shows that there is a 3% chance that the expected 7 ions may only be 2 ions observed, i.e. a signal not detectable above the amplifier noise level. Thus, as the gain is increased, the dynamic range of signal that can be recorded is decreased.

In practice, using a multiplier gain of 10^5 and assuming an instrumental sensitivity of 10^{-9} C/μg at the base peak being monitored, a peak with a maximum signal of 34 μV would result from the injection onto the GC column of 0·1 pg (10^{-13} g) with a GC peak width of 15 s and a 50% transfer efficiency of sample from GC to source. The peak itself would contain just over 300 ions. Similarly, an injection of 60 ng (6×10^{-8} g) sample would give the maximum output of 10 V. Although detection limits are usually determined by other considerations such as MS background and the ability to transfer very small samples without loss by absorption, amounts as small as 0·1 pg have been detected in practice (34). Thus, the requirement to be able to record this large dynamic range of signal is real. However, as fast A/D converters do not have such a large dynamic range, some pre-treatment of the signal or an alternative method must be considered.

8.8.1 Use of Analogue to Digital Converter Systems

Some of the earlier multiple ion monitoring systems sampled the ion signal a large number of times and then averaged these samples to provide each data point (27, 30, 31). Such systems used amplifiers of low band pass for the MS output to reduce the noise level prior to digitization and low bandpass filters are a standard feature of hardware multi-peak monitoring systems. Where such filtering of the signal is used, it must not be too efficient, otherwise traces of the signal from the previous mass will remain past the settling period and contaminate the signal from the new mass. This is particularly important when a weak peak follows a relatively intense one. The system described by Holmes *et al.* (27) allows visual examination of the ion signal in order to determine a filter value, suitable for the conditions employed. The dynamic range may be increased by dividing the MS output amongst a number of amplifiers having different gains and then allowing the A/D converter to sample each of these channels in turn through a multiplexor. The most significant output is then chosen for digitization and further processing. By this means, systems with dynamic ranges up to 500 000 have been provided. An equivalent system uses a programmable amplifier under data system control (36) (see Fig. 8.4).

Most present systems operate by integrating the ion signal and then sampling this integrated voltage with the A/D converter, the integrator being reset to zero during the switching period (27, 28, 32). A variable sampling period is usually available so that for masses where signals of low intensity are expected, longer sampling periods may be used and vice versa (27, 28, 32, 37). Such a system is able to effectively reduce the dynamic range of the incoming signals to the A/D converter but may only be used with some prior knowledge of the relative signal intensities expected at each mass. An improvement of this method is to allow the data system to sample each mass for a variable time or until the A/D converter saturates (32).

An alternative method is to employ a relatively high bandwidth for recording, e.g. 1 kHz, with direct sampling by the A/D converter. Under these conditions, at low ion currents, single ion pulses are recorded as discrete events (see Fig. 8.7). With sufficient multiplier gain and a sufficiently high sampling rate, it is then possible to ensure that each single ion pulse will be sampled more than a minimum number of times. For example, the

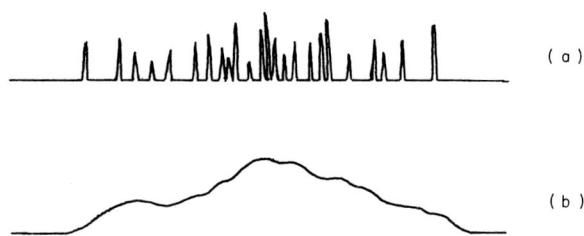

Fig. 8.7. Low ion current signal recorded (a) at high bandwidth; (b) at low bandwidth.

average height of a pulse at a gain of 2×10^5 is now approx. 5 mV (pp. 25–26), more than sufficient to ensure detection using a 12 bit A/D converter. One advantage of this method arises from the fact that the signal level will not rise above the height of a single ion pulse until more appreciable ion currents (approx. 3000 ion s^{-1} in this case) are recorded at the collector. Thus, the dynamic range required for the A/D converter may be reduced by a factor of approx. 100 from 10^6 to 10^4.

8.8.2 Use of Ion Counting Systems

A detector such as the electron multiplier may also be used in the ion counting mode (24, 38). In this mode, the output pulse from each ion incident on the multiplier is amplified by a high bandpass amplifier. These amplified output pulses meet a discriminator which rejects all those smaller than a certain size so as to reduce background noise, while passing most of the single ion pulses. Each pulse that passes the discriminator is then shaped to give an output pulse of standard amplitude and width, regardless of the variation in input amplitude. It is these standardized pulses that are then counted. Thus, ion counting systems provide a very low detector noise and allow the best possible precision in the measurement of small signals. Unfortunately, ion counting has disadvantages for general application in mass spectrometry. Counting losses occur at higher count rates because of pulse overlap, so that the system is of little value for the detection of stronger ion currents. In addition, continuous operation at relatively high electron multiplier gains is necessary in the ion counting mode.

Some practical applications of ion counting have been made in multiple ion monitoring systems. For example, Schoeller and Hayes (23) described a computer controlled ion counting isotope ratio MS applied to the measurement of ^{13}C as carbon dioxide ($^{13}CO_2$) (Fig. 8.8). For this determination, measurements at m/e 44 and m/e 45 were taken at a single collector by switching the accelerating voltage under data system control, whilst ions were detected using an electron multiplier, pulse amplifier-discriminator and a 16 bit, 300 MHz counter. The whole ratio determination procedure, including sample switching, was under computer control and, by the use of special sample handling techniques, sample sizes as small as 10^{-7} g carbon could be accommodated. An experimental precision of 1·0%, equal to that expected on the basis of counting statistics, was achieved, even at these low sample levels.

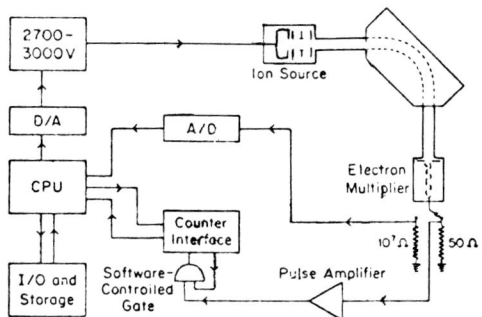

Fig. 8.8. Block diagram of the computer-controlled ion counting isotope ratio mass spectrometer. (Reprinted with permission from ref. 23. Copyright by American Chemical Society.)

Horning and coworkers (39) have described an ion counting system for use in conjunction with an atmospheric pressure ionization source and quadrupole MS. A particular advantage of the capacitatively coupled pulse amplifier–discriminator system used was that it allowed the multiplier anode to be operated at the +5 kV potential required for negative ion operation which was extensively studied with this source. Again, the majority of applications of this system were the detection of very low sample levels by single or multiple ion monitoring (39, 40). The ion detection system used by Horning comprised a Spiratron electron multiplier and pulse amplifier–discriminator. Counting used a high speed 12 bit counter to give a maximum count rate of 4×10^7 pulses/s, an improvement on an earlier system using a software program for counting which had a maximum count rate of 2×10^5 pulses/s (40).

Another use of ion counting is in the isotopic analysis of radioactive elements using a thermal ionization source. In these determinations, it is desirable to reduce sample sizes as much as possible so as to reduce contamination effects. Thus, high sensitivity becomes an important instrumental requirement. While a conventional d.c. amplifier system (Section 2.2.2) can measure ion currents down to perhaps 10^{-18} A (41), counting methods have been used in this field to measure ion currents smaller than 10^{-20} A (42, 43). In addition, recognition of signals is aided by the fact that statistical values of the measured properties are more easily determined from ion counting data than from data obtained by A/D conversion (43). A number of references to the use of data systems with thermal ionization MSs using both electron multiplier (24, 25) and scintillator detectors (26) in the ion counting mode have appeared.

8.9 DATA PROCESSING FROM SELECTED ION MONITORING EXPERIMENTS

The major application of single and multiple ion monitoring in mass spectrometry is the quantitative determination of selected compounds present in complex environments. Many data systems use a visual display unit (VDU) to present results from this type of analysis and the operator is involved interactively to a greater or lesser extent in processing these results. Thus, for inspection of peak monitoring data, typical facilities include the display of a limited section of the run, retention of only selected traces on the display and the application of gain to selected channels (e.g. Fig. 8.9).

Manipulation of the data in this way using a VDU can be controlled by means of vertical and horizontal cursors. Again, for the calculation of the area under mass peaks, a vertical cursor may be used to define integration limits and a horizontal cursor to define a background level which may vary linearly with time (Fig. 8.9). The recorded data can be smoothed using a digital smoothing routine (11) prior to area calculation to permit easier assignment of cursor positions (Fig. 8.9). Visualization of low intensity peaks is also improved by means of the smoothing routine.

For each peak chosen, peak areas, peak heights and retention times should be directly available. When dealing with more than one mass channel, interchannel peak area or height ratios should also be directly available. In theory, the measurement of peak area should give better precision than the measurement of peak height (see Appendix); however, this is not borne out in practice. One practical problem with the use of total peak area is the critical placement of the area limiting cursors, particularly in the presence of unresolved components. For example, the data in Table 8.3 shows that in a practical situation, whilst peak height measurement gives a precision of 4.0%, total area measurement can only give a precision of 4.1%. However, the use of central area measurement improves the precision to 2.7% (31). For calculation of the central area, within the cursor limits set by the operator, the computer finds the peak height, calculates the half height and locates the point at which the peak profile intersects the coordinates of half height (Fig. 8.10). Thus, the placement of cursors is no longer critical so long as other components are resolved below the half height of the major peak. Other workers have found that total peak area gives only the same or poorer precision than peak height (27, 30, 44, 45).

For routine quantitative analysis where a large number of samples are to be analysed for the same compound, for example the assay of drugs and their metabolites in biological fluids, a more automated approach to data manipulation is possible. A good example of this approach is given in a paper by Hammar et al. (28). In this system, the first defined peak and its mass channel represent the compound to be determined and the second peak and its mass channel represent the internal standard. By visual examination of one run, the peak positions for the compound to be determined and the internal standard are defined together with a search range (e.g. ± 10 scans). These search ranges are subsequently scanned for the highest intensity values to provide peak height data from other runs. Representative background positions must also be defined. Once in possession of this information, the program will automatically calculate the peak height ratio for sample to internal standard for as many runs as are required.

The program also allows the definition of information fields for each run prior to injection of the sample. Two forms of definition are valid: (a) SR, where S labels the file as a run of standards and R is the weight ratio of the compound to be determined to the internal standard: (b) $PWVT$, where P labels the file as a sample run to which W ng internal standard has been added, V ml is the sample volume which has been extracted and T stands for the time after drug administration at which the sample was taken. With this information and using the relevant standard and sample runs, the program is able to proceed automatically to tabular and graphical presentations of quantitative analyses as shown in Table 8.4 and Figs 8.11 and 8.12.

A more complex form of automatic data manipulation has been described by Summons et al. (33) for the simultaneous quantitative determination of 12 amino acids in biological fluids. The amino acids were determined as their n-butyl ester N-trifluoroacetyl derivatives with a mixture of deuterated amino acid derivatives as an internal standard. One mass for each acid and one for each deuterated acid were monitored in a cyclic manner using a computer controlled quadrupole MS, a complete cycle taking 2 s. A typical TIC trace and mass chromatograms are shown in Figs 8.13 and 8.14 respectively.

The mass chromatogram data is then reduced by a computer program requiring no operator intervention. The first processing step is to construct an approximation to the background so that it can be removed. First a "piece-wise least squares fit" through all of the relative minima is constructed

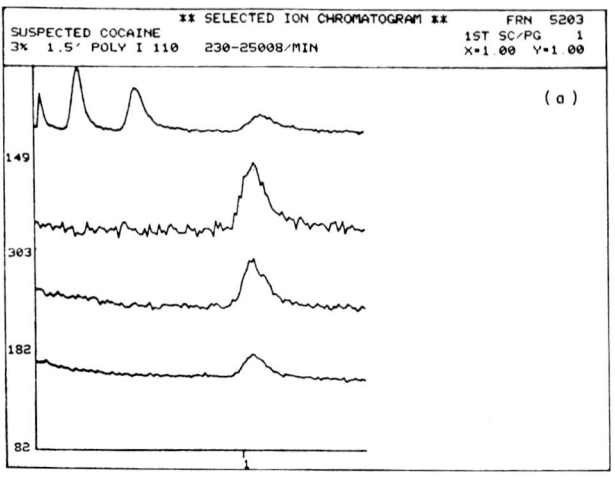

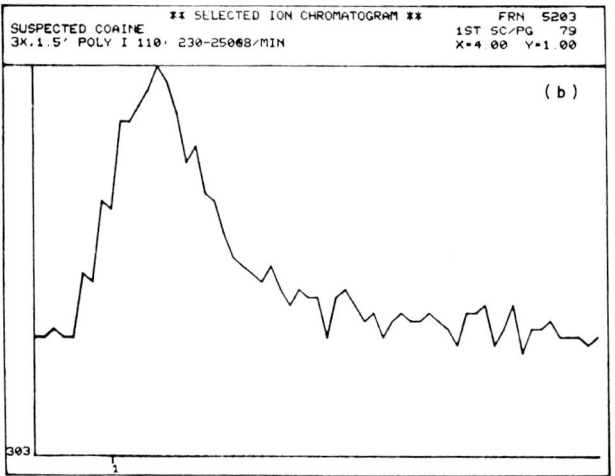

and then minima that fall above this first approximation, such as those between unresolved peaks, are eliminated prior to a second pass. Piece-wise fitting constructs local least squares fits to successive small groups of points and these solutions are then joined to give a continuous function. This procedure has produced reasonable background approximations even to quite complex chromatograms.

After subtraction of this background approximation, a threshold for peak detection is set. This is done by calculating a histogram of intensities from the background corrected mass chromatogram. Sample points are relatively

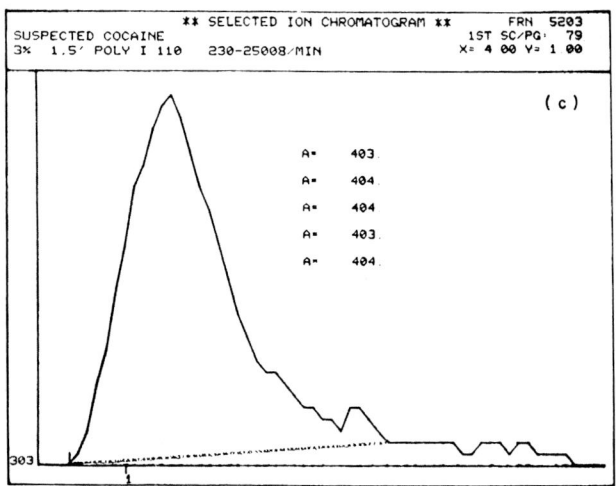

Fig. 8.9. (a) Selected ion monitoring data from the analysis of a suspected cocaine sample. (b) Isolation and expansion of trace at m/e 303. (c) Trace at m/e 303 after background subtraction and smoothing. The results of five separate area determinations and the baseline used for these determinations can also be seen. (Courtesy Hewlett-Packard.)

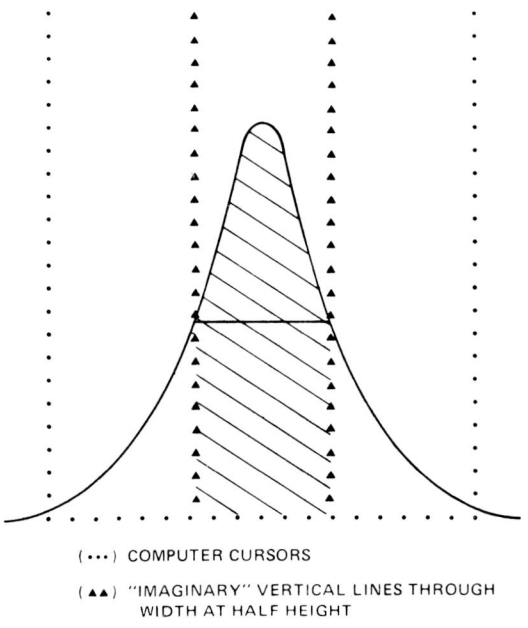

(···) COMPUTER CURSORS

(▲▲) "IMAGINARY" VERTICAL LINES THROUGH WIDTH AT HALF HEIGHT

(⧄) SHADED AREA = "CENTRAL" PEAK AREA

Fig. 8.10. Schematic representation of "central" peak area. (Reprinted with permission from ref. 31. Copyright by American Chemical Society.)

TABLE 8.3
Precision of ion monitoring results[a]

Sample no.	Central area of m/e 427	Central area of m/e 423	Ratio 427/423
1	1750	37·8	46·3
2	1691	34·2	49·4
3	1867	35·9	52·0
4	2301	43·6	52·8
5	1964	39·7	49·5
6	1976	44·5	44·4
			Mean = 49·1 ± 1·3

Summary of results for different methods of measuring same data

No. of replicates	Ratio of peak heights	Ratio of total peak areas	Ratio of central area
N = 6	50·3 ± 1·98 (4·0%)[b]	58·9 ± 2·4 (4·1%)	49·1 ± 1·3 (2·7%)

[a] Comparison of methods of calculation of low-level prostaglandin (400 pg $PGF_{2\alpha}$-ME-TMS) (m/e 423) and internal standard (20 ng d_4-$PGF_{2\alpha}$-ME-TMS) (m/e 427) injected into GCMS.
[b] Standard deviation of mean expressed as percentage of mean. (Reprinted with permission from ref. 31. Copyright by American Chemical Society.)

few and are spread over higher intensity values so that they appear as a tail on the histogram. Thus, by detecting the mode of the background histogram peak and measuring its width (standard deviation), a data adaptive threshold value can be set as: threshold = mode + 2·5 × standard deviation. Peaks were further screened by a minimum width criterion and the positions of remaining peaks determined by curve fitting about the maximum. Having searched for amino acid peaks within limits of known retention times, quantitative determination of these acids was effected by calculation of area ratios in cases where both the internal standard and sample amino acid had been located.

In other applications of MPM, for example in stable isotope ratio determination, the data manipulation required can be much more straightforward. The SIRMID unit described by Klein *et al.* (22) is a case in point. Once the masses at which determinations are to be made have been set up, background is counted for each mass and these values transferred to the memory. When the ion intensity due to sample begins to increase, the operation mode is switched to start. Now the counts accumulated during the sweep across each mass peak are stored in a register and shown in two LED displays. A third display gives the isotope ratio directly. This ratio is

TABLE 8.4

One analysis of a patient's blood level after a single oral dose of 100 mg of Kabi 1774[a]

100 mg of KABI 1774 Patient no. 5		
Range = 7	Chan. = 1	Background = 250
Peak 1 = 430	Time = 4·83	
Peak 2 = 495	Time = 5·56	

731013
Standard files

File	W ratio	P ratio
47	0·400	0·462
48	0·800	0·870
49	1·510	1·506
50	2·520	2·536

731013
Analysis results

File	Concentration	Time
51	619·13	2·00
52	1204·27	3·00
53	2656·38	4·00
54	3674·08	6·00
55	1717·95	8·00
56	1193·69	10·00
57	254·76	24·00
58	208·04	28·00
59	156·76	32·00
60	102·22	48·00
61	67·15	56·00
62	42·97	72·00

[a] Concentration is expressed in ng/ml serum, and the time is the number of elapsed hours between the taken dose and the drawn blood sample. (From ref. 28.)

calculated from the background corrected counts in each channel and is continuously updated.

Another low cost data acquisition and analysis system for isotope analysis has been described by Hulston et al. (46). This system, based on a programmable, printing calculator acquires data from a double collector MS and calculates the small variations of natural carbon, oxygen and sulphur isotope ratios that can be determined in this way.

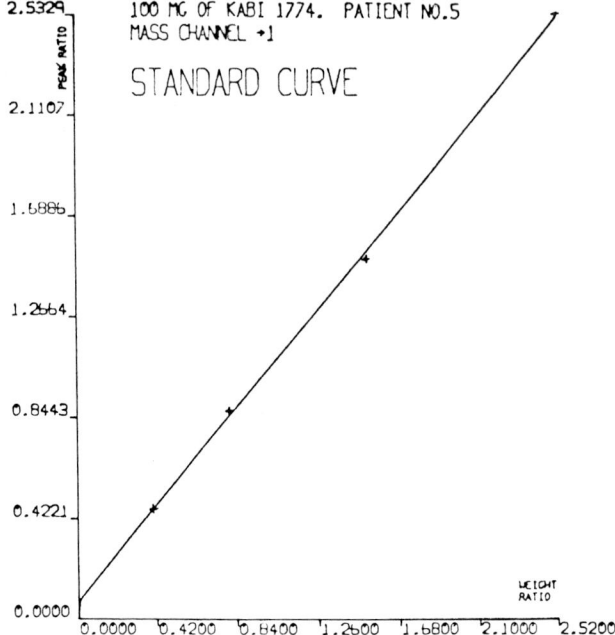

Fig. 8.11. A computer determined and plotted standard curve based upon selected ion monitoring data from four analyses of different weight ratios between the compound to be determined and the internal standard. (From ref. 28.)

A review of the statistical considerations involved in the reduction of measured ion intensities, together with calibration data, to give the quantity of a labelled species or of a particular compound present, has recently been presented by Schoeller (47).

8.10 QUANTITATIVE DATA FROM SPARK SOURCE ANALYSES

Spark source mass spectrometry is generally considered to be one of the most sensitive and comprehensive techniques for the determination of trace elements in solids. The spark source is used on instruments of Mattauch–Herzog geometry generally with an ion sensitive photographic plate detector which is capable of integrating the entire mass spectrum simultaneously and is therefore well suited for trace survey analysis. Unfortunately, the blackening of a photographic plate shows a complex dependence on incident ion intensity and on ion mass. Thus, the published computer-based procedures used to transform digitized transmittance data read from the plate

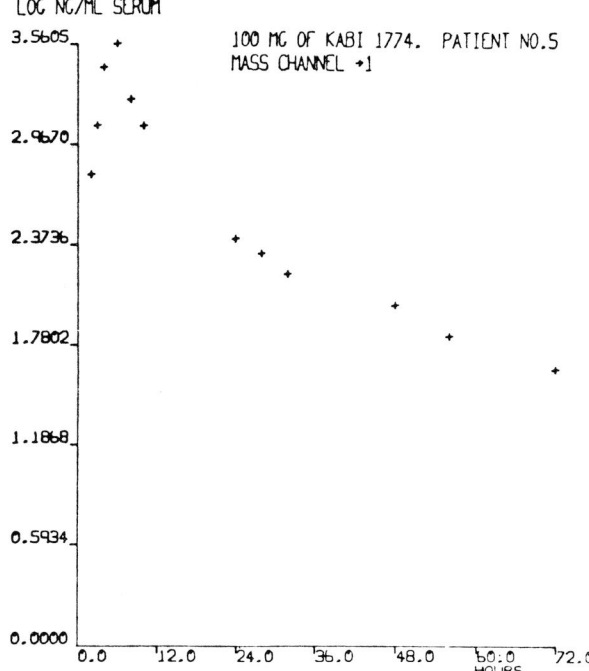

Fig. 8.12. A computer determined and plotted dose response curve. The blood concentration of the drug in a patient, at various times after the single oral dose is determined by means of the standard curve (Fig. 8.11) and the selected ion monitoring data from the analyses of the serum extracts. (From ref. 28.)

(Section 2.5) into ion intensity are complex and differ considerably in detail among themselves (48–50). However, a very brief summary is possible and the reader is then directed to the excellent papers describing these procedures. A number of authors also give very full details of the procedures used for the computer-based assignment of ion compositions in spark source spectra (48, 49, 51).

Most authors have used the Hull equation (52) (eqn. 8.9) to describe the relationship between incident ion intensity (E) and the experimentally recorded transmittance (T):

$$E = 1/K_0((1 - T)/(T - T_s))^{1/G}. \qquad (8.9)$$

Here, K_0 is a photographic sensitivity constant, G is a constant dependent on the slope of the characteristic curve and T_s is the saturation transmittance. The constants G and K_0 are determined from a fit of log E vs. log$(1 - T)/(T - T_s)$ for a number of calibration elements. However, it is necessary to correct

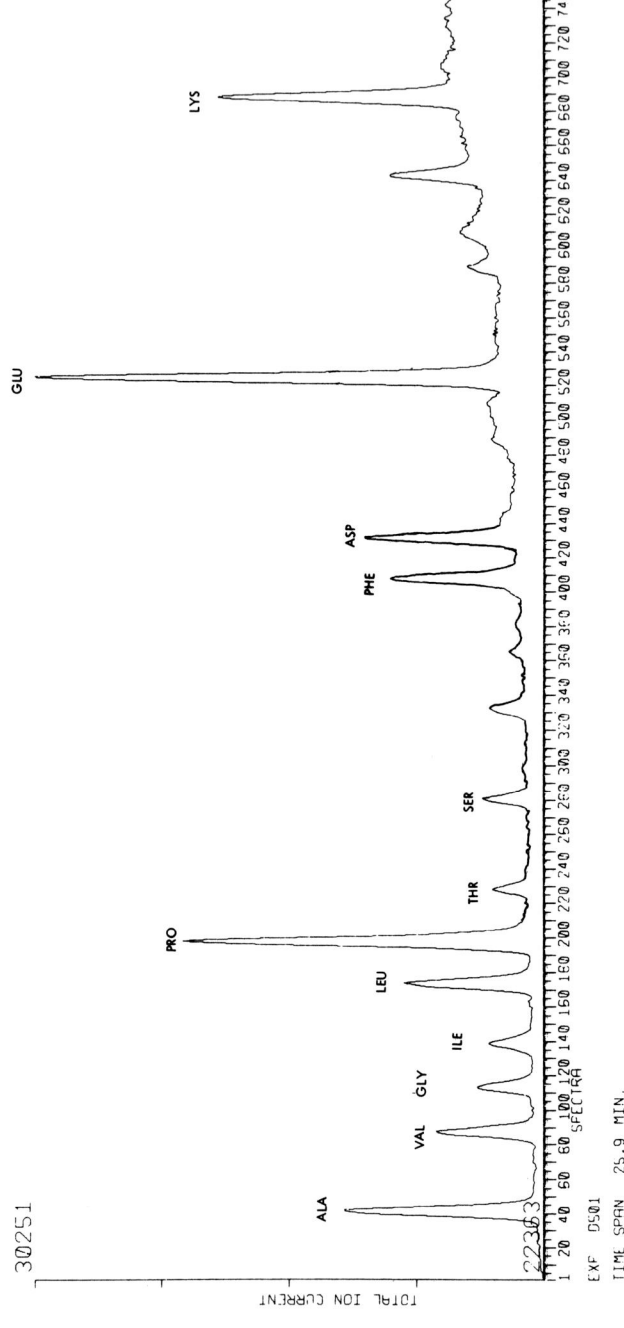

Fig. 8.13. TIC from a normal plasma. (Reprinted with permission from ref. 33. Copyright by American Chemical Society.)

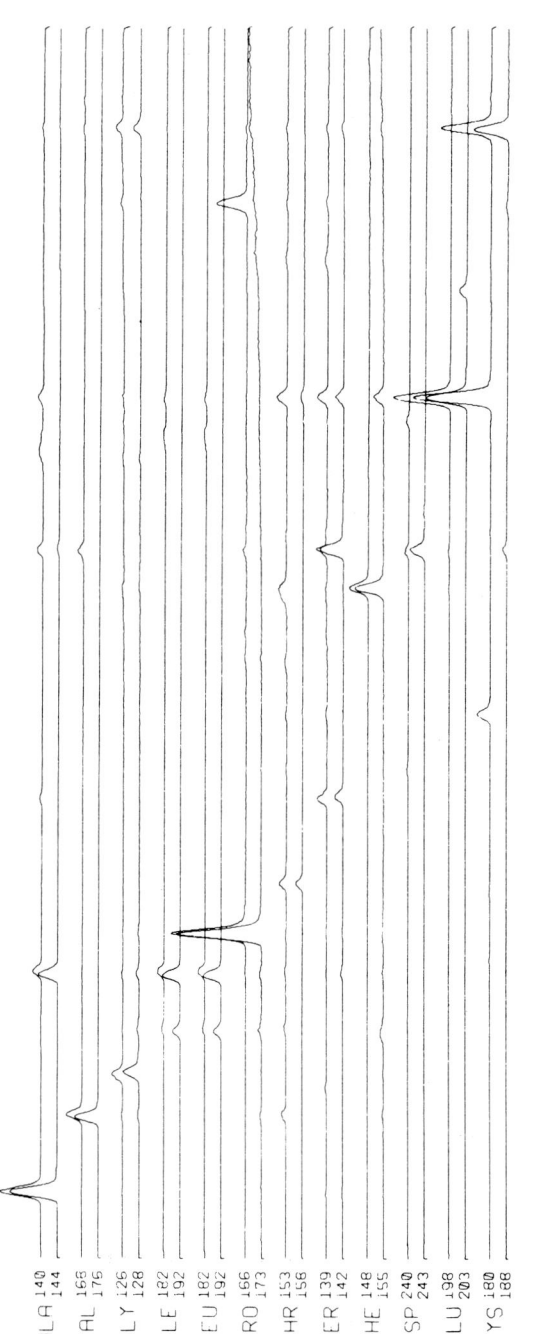

Fig. 8.14. Individual ion chromatograms of monitored fragments. (Reprinted with permission from ref. 33. Copyright by American Chemical Society.)

these constants when used for other elements as they are both mass dependent. The saturation transmittance (T_s), is measured from saturated lines produced by a calibration element, but this factor too must be corrected for mass dependence. In addition, all transmittance values (T) used must be corrected for the background contribution at this point. The concentration of each element is finally determined with respect to a standard or matrix element from the appropriate integrated line intensities (Table 8.5). Procedures for the determination of ion intensities from the spectra of organic materials recorded on photoplates have received much more limited consideration than those for spark source spectra. However, details of some procedures are available (53–55).

TABLE 8.5

Interpretation of copper in a mild steel matrix [a]

Extract from line list		
Mass	Area	Possibles
30·971	195·6	P 52·5
31·415	11·0	
31·465	33·5	
31·969	274·9	S 78·8
32·462	19·2	
33·964	15·1	S 100
62·930	522·7	Cu 289
63·930	44·9	Ni 1600 Zn 35·3
64·929	218·2	Cu 274

63 agrees with 65

31·5 agrees with 63 within 8/1 to 200/1
32·5 agrees with 65 within 8/1 to 200/1

∴ Element confirmed at concentration = 281 p.p.m.

[a] Confirmation is by agreement between the two isotopes and location of the doubly-charged ions. (From ref. 49.)

Owing to the problems associated with photoplate detection and data reduction from photoplates, a number of authors have experimented with electrical detection systems for spark source work, mostly in the peak switching mode. For example, Morrison *et al.* (56) have described an electrical detection system in which an on-line computer controls both the electrostatic analyser (ESA) and the magnet; the latter either via magnet current or via the magnetic field sensed by a Hall probe. In order to locate masses for peak switching analysis, the computer sets the magnet field so that the required mass occurs in the m to $2m$ mass region as defined by the range of the ESA voltage. The computer then scans the ESA voltage and acquires data from this scan. After the operator has assigned a mass to at least one point, a mass to ESA voltage relationship is calculated and the precise masses selected by the operator set up for peak switching analysis. Very rapid quantitative analyses are possible with on-line data acquisition.

REFERENCES

1. Sweeley C. C., Elliott W. H., Fries I. and Ryhage R. (1966). *Analyt. Chem.* **38**, 1549.
2. Hood A. and O'Neal M. J. (1959). *In* "Advances in Mass Spectrometry" (Ed. Waldron J. D.), Vol. 1, pp. 179–192. Pergamon Press, London.
3. Robinson C. J. and Cook G. L. (1969) *Analyt. Chem.* **41**, 1548.
4. Robinson C. J. (1971) *Analyt. Chem.* **43**, 1425.
5. Gallegos E. J., Green J. W., Lindeman L. P., Le Tourneau R. L. and Teeter R. M. (1967). *Analyt. Chem.* **39**, 1833.
6. Fisher I. P. and Fischer H. P. (1973). 21st Annual Conference on Mass Spectrometry and Allied Topics, San Francisco. Paper U7, p. 492.
7. Barnard G. P. (1953). "Modern Mass Spectrometry". Institute of Physics, London.
8. Ruth P. (1968) *Analyt. Chem.* **40**, 747.
9. Seppä I. and Multala R. (1973). *Kemian Teollisuus* **30**, 457.
10. Raimondi D. L., Winter H. F., Grant P. M. and Clarke D. C. (1971). *IBM J. Res. Develop.* **15**, 307.
11. Savitzky A. and Golay M. J. E. (1964). *Analyt. Chem.* **36**, 1627.
12. Tunnicliff D. D. and Wadsworth P. A. (1965). *Analyt. Chem.* **37**, 1082.
13. Tunnicliff D. D. and Wadsworth P. A. (1973). *Analyt. Chem.* **45**, 12.
14. (a) Axelson M., Cronholm T., Curstedt T., Reimendal R. and Sjövall J. (1974). *Chromatographia*, **7**, 502.
    (b) Sjövall J. (1975). *J. Steroid Biochem.* **6**, 1.
15. Baczynskyj L., Duchamp D. H., Zieserl J. F. Jr. and Axen U. (1973). *Analyt. Chem.* **45**, 479.
16. Schuetzle D., Crittenden A. L. and Charlson R. J. (1973). *J. Air Pollution Control Ass.* **23**, 704.
17. Rosenthal D. and Bursey J. T. (1972). 20th Annual Conference on Mass Spectrometry and Allied Topics, Dallas. Paper T4, p. 419.
18. Dowie R. S., Kemball C., Kempling J. C. and Whan D. A. (1972). *Proc. R. Soc. Lond.* (A) **327**, 491.
19. Haddon W. F., Lukens H. C. and Elsken R. H. (1973). *Analyt. Chem.* **45**, 682.
20. Chapman J. R., Evans S. and Wolstenholme W. A. (1970). 18th Annual Conference on Mass Spectrometry and Allied Topics, San Francisco. Paper B11, p. B 42.

21. Young N. D., Holland J. F., Gerber J. N. and Sweeley C. C. (1975). *Analyt. Chem.* **47**, 2373.
22. Klein P. D., Haumann J. R. and Hachey D. L. (1975) *Clin. Chem.* **21**, 1253.
23. Schoeller D. A. and Hayes J. M. (1975). *Analyt. Chem.* **47**, 408.
24. Ihle H. R. and Neubert A. (1971). *Int. J. Mass Spectrom. Ion Phys.* **7**, 189.
25. Christie W. H. and McKown H. S. (1971). *Chem. Instrum.* **3**, 99.
26. Lagergren C. R. and Stoffels J. J. (1970). *Int. J. Mass Spectrom. Ion Phys.* **3**, 429.
27. Holmes W. F., Holland W. H., Shore B. L., Bier D. M. and Sherman W. R. (1973). *Analyt. Chem.* **45**, 2063.
28. Hammar C-G., Pettersson G. and Carpenter P. T. *Biomed. Mass Spectrom.***1**, 397.
29. Klein P. D., Haumann J. R. and Eisler W. J. (1972). *Analyt. Chem.* **44**, 490.
30. Holland J. F., Sweeley C. C., Thrush R. E., Teets R. E. and Bieber M. A. (1973). *Analyt. Chem.* **45**, 308.
31. Watson J. T., Pelster D. R., Sweetman B. J., Frolich J. C. and Oates J. A. (1973). *Analyt. Chem.* **45**, 2071.
32. Caprioli R. M., Fies W. F. and Story M. S. (1974). *Analyt. Chem.* **46**, 453A.
33. Summons R. E., Pereira W. E., Reynolds W. E., Rindfleisch T. C. and Duffield A. M. (1974). *Analyt. Chem.* **46**, 582.
34. Wilson B. W., Parker R. B. and Snedden W. e.g. (1976). *In* "Advances on Mass Spectrometry in Biochemistry and Medicine", Vol. 2. Spectrum Publications, New York.
35. Sweeley C. C., Young N. D., Holland J. F., Gates S. C. (1974). *J. Chromatog.* **99**, 507.
36. Elkin K., Pierrou L., Ahlborg U. G., Homstedt B. and Lindgren J-E. (1973). *J. Chromatog.* **81**, 47.
37. Eichelberger J. W. and Budde W. L. (1974). 22nd Annual Conference on Mass Spectrometry and Allied Topics, Philadelphia. Paper D6, p. 129.
38. White F. A. and Collins T. L. (1954). *Appl. Spectrosc.* **8**, 17.
39. Carroll D. I., Dzidic I., Stillwell R. N., Haegele K. D. and Horning E. C. (1975) *Analyt. Chem.* **47**, 2369.
40. Horning E. C., Horning M. G., Carroll D. I., Dzidic I. and Stillwell R. N. (1973). *Analyt. Chem.* **45**, 936.
41. Beynon J. H. (1960). "Mass Spectrometry and its Applications to Organic Chemistry", Elsevier, Amsterdam.
42. Werner H. W. and de Grefte H. A. M. (1965). Trans. Third Int. Vac. Congr., Vol. 2, p. 493. Pergamon Press, New York.
43. Blauth E. W., Draeger W. M., Kirschner J., Liebl H., Müller N. and Taglauer E. (1971). *J. Vac. Sci. and Technol.* **8**, 384.
44. Alford A. L. (June 1974). EPA report No. EPA-660/2-74-002, "Evaluation of a computer program for GC-MS specific ion monitoring".
45. Holland P. T. (January 1975). Third Biennial Conference of the Australian and New Zealand Society for Mass Spectrometry. Canberra.
46. Hulston J. R., Low K. W. and Blomfield T. G. (1973). *Control and Instrument*, **5**, 48.
47. Schoeller D. A. (1976). *Biomed. Mass Spectrom.* **3**, 265.
48. Frisch M. A. and Reuter W. (1973). *Analyt. Chem.* **45**, 1889.
49. Millett E. J., Morice J. A. and Clegg J. B. (1974). *Int. J. Mass Spectrom. Ion Phys.* **13**, 1.
50. Burdo R. A., Roth J. R. and Morrison G. H. (1974). *Analyt. Chem.* **46**, 701.
51. Brown R., Powers P. and Wolstenholme W. A. (1971). *Analyt. Chem.* **43**, 1079.

52. Hull C. W., 10th Annual Conference on Mass Spectrometry (ASTM Committee E14), New Orleans, p. 404.
53. Desiderio D. M. (1971). *In* "Mass Spectrometry, Techniques and Applications", (Ed. Milne G. W. A.), p. 11. John Wiley and Sons, Chichester.
54. Venkataraghavan R., Board R. D., Klimowski R., Amy J. W. and McLafferty F. W. (1968). *In* "Advances in Mass Spectrometry" (Ed. Kenrick, E.), Vol. 4, p. 65. Institute of Petroleum, London.
55. Habfast K. (1968). *In* "Advances in Mass Spectrometry" (Ed. Kendrick E.), Vol. 4, p. 3. Institute of Petroleum, London.
56. Morrison G. H., Colby B. N. and Roth J. R. (1972). *Analyt. Chem.* **44**, 1203.

Appendix

USEFUL FORMULAE AND DEFINITIONS

Mass spectrometer sensitivity: The basic sensitivity of a mass spectrometer may be defined as the total charge collected at a specified mass during the consumption of a known amount of sample in the source.

For example, a typical figure is:
10^{-9} Coulomb (C) at m/e 74 (base peak of the spectrum) per microgram of methyl stearate.

Charge on a single ion: Charge on a single ion = 1.6×10^{-19} C.

Number of ions in a scanned mass spectral peak: Let mass spectral peak width = t_p s (see below), recorded peak height = V volts, amplifier input resistor (p. 21) = R_f Ω, multiplier gain = G, sensitivity at mass concerned = S C µg^{-1}, sample flow rate into source = Q µg s^{-1} and assume the mass spectral peak has a triangular profile, then:

$$\text{Area} \equiv \tfrac{1}{2} \frac{Vt_p}{R_f G} \times \frac{1}{1.6 \times 10^{-19}} \text{ ions.}$$

(this equation provides a convenient means of determining multiplier gain from the recorded area of a single ion peak). Alternatively, the number of ions in the peak is given by:

$$\tfrac{1}{2} \times \frac{SQt_p}{1.6 \times 10^{-19}}.$$

Mass spectral peak width in exponential magnet scanning: Let time to scan one decade in mass = t_{10} s, resolving power (based on 10% valley definition) = R, peak width at 5% height = t_p s, then:

$$t_p = \frac{t_{10}}{2.303\,R}$$

Probability of observation of peaks containing small numbers of ions: If the expected number of ions in a mass spectral peak is N, then the probability of observing an actual number k in a particular scan is the appropriate term in the expansion:

$$e^{-N}\left(1 + N + \frac{N^2}{2!} + \frac{N^3}{3!} + \ldots + \frac{N^k}{k!} + \ldots\right)$$

$$\text{i.e. } P(k) = \frac{e^{-N} \cdot N^k}{k!}.$$

This is the Poisson distribution, which applies to small values of N. The total probability is unity, since $(1 + N + N^2/2! + \ldots) = e^N$.

When N becomes small (less than 20) there is a significant chance that the peak observed in an individual scan will be much smaller than expected, or even non-existent. For example, if $N = 1$, there is a very significant probability of observing no ions at all ($P(0) = 0.37$).

In general, the probability of observing a number of ions less than or equal to a number m is:

$$P(\leqslant m) = e^{-N}\left(1 + \sum_{k=1}^{m} \frac{N^k}{k!}\right).$$

Values of $P(\leqslant m)$ are given in the table overleaf.

Variation in recorded peak position and therefore mass measurement: Let number of ions in the peak $= N$, resolving power $= R$, then, standard deviation of peak position, determined as the centroid position is, given by:

$$\sigma = \frac{10^6}{R\sqrt{24N}}$$

Statistical variation in peak area for larger peaks: Let the peak contain N ions, then, relative standard deviation of peak area (A) determination is given by:

$$\frac{\sigma_A}{A} = \frac{1}{\sqrt{N}}$$

Statistical variation in recorded signal level: Let ion current at collector $= n$ ions s^{-1}, bandwidth of recording system $= f$ Hz (time constant $= \tau = 1/2\pi f$ s), then, relative standard deviation of signal level (H) determination (e.g. peak height or constant level) is given by:

$$\frac{\sigma_n}{H} = \sqrt{\frac{\pi f}{n}} = \sqrt{\frac{1}{2n\tau}}.$$

Values of $P(\leq m)$ for $N = 1$ to 10

N^a	0^b	1	2	3	4	5	6	7	8	9	10
1	0·368	0·736	0·920	0·981	0·996	1·000	1·000	1·000	1·000	1·000	1·000
2	0·135	0·406	0·677	0·857	0·947	0·984	0·996	0·999	1·000	1·000	1·000
3	0·050	0·199	0·423	0·647	0·815	0·916	0·967	0·988	0·996	0·999	1·000
4	0·018	0·092	0·238	0·434	0·629	0·785	0·889	0·949	0·979	0·992	0·997
5	0·007	0·041	0·125	0·265	0·441	0·616	0·762	0·867	0·932	0·968	0·986
6	0·003	0·017	0·068	0·151	0·285	0·446	0·606	0·744	0·847	0·916	0·957
7	0·001	0·007	0·030	0·082	0·173	0·301	0·450	0·599	0·729	0·831	0·902
8	0·000	0·003	0·014	0·042	0·100	0·191	0·313	0·453	0·593	0·717	0·816
9	0·000	0·001	0·006	0·021	0·055	0·116	0·207	0·324	0·456	0·588	0·706
10	0·000	0·001	0·003	0·010	0·029	0·067	0·130	0·220	0·333	0·458	0·583

[a] 0 to 10 (across) are values of m.
[b] 1 to 10 (down) are values of N.

With a background level equivalent to x times the signal level, then relative standard deviation is given by:

$$\frac{\sigma_n}{H} = \sqrt{\frac{\pi f(1 + x)}{n}}.$$

Subject Index

A

Abbreviation of spectra, 106–120
   reduction of storage requirements, 106
   reduction of variability, 106, 127
Accelerating voltage alternator (AVA), 229
Accurate mass data quantitative analysis, 224, 228
Acetylaminoethylphosphonic acid, TMS ether, 115
Acetylcholine, deuterated, 90
Air pollutant analysis, 224–225
Alkylbenzenes, 152–153, 162–163
Allbreaks, 209
Amines, 201
Amino acids, 241–244
Analogue to digital converter
   continuous operation, 23–27
   discontinuous operation, 36–39
   use with sample and hold amplifier, 27
   sampling rate, 27–28
   setting of sampling rate, 29
   specification for scanning, 25–26
   specification for selected ion monitoring, 237
   successive approximation, 23–25
   variable sampling period, 237
Anomalous component location, 144
Archaeological material, source identification, 182
Arpanet, 213
Atmospheric pressure ionization, 239
Autoscaling, 173–174
Averaging of scans, 64, 226–227

B

Background spectra
   automatic choice, 82
   subtraction, 82
Bacteria, classification, 180
Badlist, 201
Basic factors, 182–183
Bit string comparison, 120, 122
Branching tree array, 166, 188, 189
Bromine, test for presence, 197

C

Calibration
   accurate mass, 53–55
   nominal mass, 46–52
Carrick system, 36–38
CASAC, 144
Central area, 240–241
Central processing unit (CPU), 9–11
Centroid, 34
   standard deviation of position, 57–58
$C_{10}H_{14}$ isomers, 182–184
Charge
   localization, 210
   single ion, 254
Chemist-learning machine interaction, 162–163
Chlorine, test for presence, 197
Cholesterol, 110–111
Chromatogram (see Homologous series chromatogram, Mass chromatogram, Mass resolved chromatogram, Mass subset chromatogram, Total ion current chromatogram)
Classification
   factor analysis, 182–184
   tests automatically devised, 191–192
Clock, 29
Cluster
   overlap, 152
   tightness, 151–152, 157

Cluster analysis, 151, 179–182
Comparator, 30, 31
Comparison routines, 120–139
  absolute intensity data, 125–131, 135
  intensity ranking, 123–125
  intensity ratios, 128, 135
  logical operator, 120–123
  optimization, 122
Component spectra location in GCMS, 81–90
Component vector, 157
Congen, 208
Continuous scanning, 77–78
Continuous valued property, prediction of, 166–167
Cortisone, 216
Covariance analysis, 113, 182
Cross correlation, 33
Creatinine, 226–228
Cross terms, 164
Cursors, 240–241
Cyphernet, 3, 129

D

Data acquisition
  focal plane mass spectrometer, 40–43
  quadrupole mass spectrometer, 39, 40
  scanning mass spectrometer, 18–39
Data bases, 103–105
  quality of data, 103–105
Data transfer, 11–12
  autonomous, 12
  program control, 11–12
  interrupt control, 12
DDT, 104
Dead-zone (*see also* rejections), 164–165
Deconvolution, 34
Dedicated system, 1, 18–19
Defocusing methods, data acquisition, 71–73
Dendral, 3, 199–214, 215, 217
Digital learning net, 167–168
Digital smoothing routine, 32, 43, 240
Digital to analogue converter, specification in multiple peak monitoring, 232

Digitization rate
  choice, 28
  effect on mass measurement accuracy, 60–62
Direct memory access (DMA), 12, 28
Disc, magnetic, 13–14
  performance data, 14
Dissimilarity index, 122, 125, 127, 131
Distance, measures, 151
Double bond equivalent (DBE), 195, 198
  prediction, 132
Drugs, 241
  activity related to MS, 182
Dynamic range of signals
  scanning, 26
  selected ion monitoring, 236–237

E

Eight Peak Index, 103, 136
Electron book-keeping, 210
Electron multiplier, 21
Elemental compositions
  from accurate mass data, 95–99
  calculation routines, 96–97
  limitation of number, 98–99
  number of possible, 98
  from learning machines, 166
  from nominal mass data, 91–95
  use with library search data, 145–146
Encoding
  absolute intensity data, 119–120
  binary, 119
  as features, 114–116
  by information content, 112–113
  as mathematical function, 116–118
  N every M mass units, 108–112
  N-most intense peaks, 107
  N-most significant peaks, 112, 125, 135
  octal, 109
  relative intensity data, 118–119
Estrogens, 202–204
External reference, mass measurement, 63
Extrapolation formulae, 54

F

Factor analysis, 182–184
   in mixture analysis, 90–91
Fatty acid methyl esters, 187, 190
Feature extraction, 160–162
   variance based, 162
Filter (*see* Pre-search)
Fisher ratio, 160, 178
Focal plane detector, direct reading, 43
Fomblin, 51
Forward search, 101
Fourier transform, 159
Fractional mass correction, 47–49
Fragmentation rules, formalization, 187, 209–213
Functional groups, classification, 152

G

Gas chromatograph retention time, 142–144
Gasoline analysis, 167, 221–223
Gaussian profile fitting, 69, 227
Goodlist, 199

H

Hadamard transform, 159
Hall probe
   data acquisition control, 38
   nominal mass calibration, 51–52
   sampling of signal, 23
Heroin, 138
Heuristics, 199
Hexadecane, 31
Hexa-2,5-dione, 84
Hierarchical system, 2
High mass
   operation at, 51
   reference for, 51
Hites-Biemann correction, 48–49
Homologous series chromatogram, 84
Hull equation, 247
Hydrocarbons, 90, 117–118, 162, 187
Hyperspace, 151, 157
Hysterisis, 59

I

Incremental recording system, 38–39
Infra-red data, joint use, 160, 174
Information gain, 155
Inorganics, quantitative analysis, 224
Input-output devices, 14
Integration of ion signal, 237
Intensity ratio, use in classifications, 176–179
Intensity transformation, 159
Interface, mass spectrometer-computer, 9
   control functions, 29–30
Interpretation
   high resolution spectra, 194–198, 202–204
      interactive scheme, 197–198
   low resolution spectra, 187–192, 199–202
Interpolation formulae, 55–57
Interrupt, 12, 29
Intsum, 209–210, 213, 216
Inverted file, 102, 129
Ion generator, 209, 210–213, 216
Ion counting, 238–240
Ion series, 140, 152–155, 187, 188
Ion statistics, mass measurement error, 57–59
Ion types, 198
Ionization efficiency data, 73
Isomers, number of, 298
Isotope pattern, calculation, 91–94
Isotopically labelled materials
   determination of radioactive, 239–240
   determination of stable, 225–228, 238–239, 244–246
   separation by gas chromatography, 225

K

Ketones, 195, 199–201
K-nearest neighbour (KNN) method, 170–174
   comparison with learning machine, 171–172

L

Language
  assembly, 15
  high level, 15–16
  machine, 15
Large scale integration (LSI), 2
Learning machines, 157–170
  comparison with KNN method, 171–172
  training, 157–158
Least squares fit
  in extrapolation, 54
  in interpolation, 55–57, 59
Least squares procedure, 69, 94, 166–167, 221, 222
Library search
  interpretation using elemental compositions, 145–146
  interpretation using GC data, 142–143
  location of anomalous components, 144
Limited mass data
  library search, 108
  structural information, 117–118
Limonene, 78–79
Linearly inseparable data, 164–165
  chemist-learning machine approach,
  unreliability of learning machine, 164
Lisp, 213
"Logical and" (AND), 121
"Logical exclusive or" (XOR), 121
Lookup table, 47, 54
Losses, structural information from, 116
Low electron voltage data, 204
Low mass ions, structural information from, 115

M

Magnet
  drift, 230–231
  switching, 232–233
Mass chromatogram 84
  background approximation, 241–242
  peak detection threshold, 242–244
Mass measurement
  low resolution, 63–64
  photoplates, 66–67
  scans, 52–64
Mass measurement error, 57–64
  accuracies of measurement and calculation, 59–60
  effect of digitization frequency, 60–62
  effect of instrumental stability, 59
  effect of ion statistics, 57–59
  effect of resolving power, 62–63
  effect of scan rate, 62–63
  reduction, 64
  units, 95–96
Mass resolved chromatogram, 86
Mass spectrometer, 3–8
  double beam, 52
    low resolution mass measurement, 63
  double focusing, 5–6
    accurate mass measurement, 52–53
  focal plane, 6
  magnetic sector
    control for selected ion monitoring, 229–235
  quadrupole, 7
    calibration, 52
    computer control, 40
    control for selected ion monitoring, 235–236
  sensitivity, 254
  single focusing, 3–5
  time of flight, 8
    calibration, 50–51
Mass Spectrometry Data Centre (MSDC), 103, 108, 124
Mass subset chromatogram, 84
Massmatch, 124, 139
Matching index (see Similarity index)
Matrix analysis, 221
Memory
  core, 12–13
    performance data, 14
  read only, 13, 15
Metastables (see Peaks metastable)
Methyl-n-propyl-n-butylamine, 212
Methyl stearate, 254
Microanalysis, 95
Microcomputer, 2
Microdensitometer, 41–42
Microprocessor, 2
Minicomputer, 1

SUBJECT INDEX

Minimum distance classifier, 151, 188
    shift of decision surface, 153
Mixtures
    quantitative analysis without separation, 220–223
    resolution of spectra, 84–91
Molecular ion location, 188, 189, 195, 204–205
Molecular weight, 139–140
    estimation, 189
Molion, 205
Moments, 159, 172
Monoisotopic spectra, 94
Monosaccharides, 176–179
MSSS, 3, 102, 103, 116, 129–131, 139–140
Multicategory prediction, 166–167
    KNN method, 167, 171
    by majority vote, 166
Multiple peak monitoring (MPM), 4, 26, 65, 228–229
Multiplets (see Peaks multiplet)
Multiplexor, 23, 237

N

^{15}N measurement, 225–228
N/D ratio, 159, 164
Networks, 3
Nitrogen, test for presence, 197
Noise (see Peaks noise)
Non-linear mapping (NLM), 180–182
Non-linear transform, 152–153
Non-parametric methods, 150
Normalization of mass spectra, 79

O

Off-line processing, 1, 19–20
Oligonucleotides, 174
On-line processing, 1, 18–19
Optimum variables, 159, 172–173
Oxygen, test for presence, 197

P

Parallel array, 166
Parametric methods, 150, 174–179
    two step method, 176

Partial factors, 182–183
Peak area, 35–36
    comparison with peak height, 240
    measurement, 240–241
    variation, 255
Peak height, 240
Peak intensity, calculation, 35–36
Peak matching, 65–66
    low resolution, 65
Peak position
    calculation, 34–36, 43
    variation, 255
Peak width, 254
Peaks
    digital smoothing, 32, 43
    mass deficient
        accurate mass measurement, 53–54
        nominal mass measurement, 39, 47–48
    metastable
        defocusing methods, 71–73, 215–216
        interpretative schemes, 202, 214–216
        recognition, 31, 69–71
    multiplet
        common doublets listed, 67
        deconvolution, 69
        recognition, 31, 67–69
    noise, 31, 33
    normal, 30–34
    number of ions, 254–255
    single ion, 31
    small, 254–255
Peptide sequencing
    high resolution data, 192–193, 194
    low resolution data, 193–194
Perfluorokerosene (PFK), 46, 51, 53, 232
Perfumery mixture, 125, 127
Phosphonates, 161–162
Photoplate, 40–41
    mass measurement, 66–67
    quantitative data, 246–250
Planner, 202
Plotting of mass spectra, 78–79
Poisson distribution, 255
Polychlorinated biphenyls (PCBs), 84
Pre-amplifier, 21–22
Precision, 102
Predictive ability, 155
    correlation with molecular complexity, 197

Predictive ability—*continued*
   statistical fluctuation, 155
Preliminary inference maker, 199–201
Preprocessing of data, 152–155, 159–162
Pre-search, 139–144
   GC retention time, 142, 144
   hurdle, 141
   ion series, 140
   mass coincidence, 139
   molecular weight, 139–140
   in reverse search, 141–142
Presence of structure method, 162
Primitive mechanisms, 210–213
Principal factors, 182
Probability based matching (PBM), 3, 112, 114, 135–139, 142
   hardware programs, 138–139
Probability distribution parameters, 174–175
Programmable calculator, 245
Purity index, 125–126

Q

Quality control, 144

R

Recall, 102
Reconstructed mass spectrum, 85
Rejections (*see also* Dead-zone), 156
Reliability, 102, 155
Remaining pot, 207
Resolution improvement, 34
Reverse search, 101, 112, 125, 135
Ripple, 59

S

Saturated acyclic monofunctional compounds (SAM), 201–202
Scan data for quantitative analyses, 220–228
   calibration, 224
Selected ion monitoring, 220, 228–246
   automatic focusing routine, 233
   bandwidth used, 237–238
   data processing from, 240–246
   dynamic range of signals, 236–237
   high resolution, 231
   mass range available, 231–232
Sequential search, 102
Serial file, 102
Series displacement index, 140
Shortest spanning path (SSP), 179–180
Signal
   conditioning, 22
   transmission, 23
   variation in level, 255–257
Silicon, test for presence, 197
Similarity index, 124, 128
Single ion, charge on, 254
SIRMID, 235, 244
SLAM, 167
Software, 15–16
Spark source mass spectrometry, 182
   electrical detection system, 250–251
   quantitative data, 246–251
Stability of instrument
   effect on accurate mass measurement, 59
   effect on nominal mass measurement, 52
   importance in selected ion monitoring, 230
Statistical occurrence of mass and abundance values, 136–137
Stepwise regression method, 222
Steroids, 82, 89, 154–155, 156
STIRS, 3, 109, 114–116, 131–135, 147, 150
Structural code, 188
Structure drawing program, 188
Structure generator, 199–201, 205–209
Substructure searching, 131–134
Sulphur
   substitution for oxygen in aliphatic compounds, 118
   test for presence, 197
Sumex-Aim, 3, 213–214
Superatom, 201, 202, 207, 208
Superatompot, 207
Supervised learning, 151

T

Tabulation of mass spectra, 78
Tape, magnetic, 13
  analogue, 19–20
  digital, 20
  performance data, 14, 20
Thermal ionization, 239
Threshold, digital, 30
Threshold logic unit (TLU), 157
Time-sharing system, 2, 19
  maintenance of library on, 3, 102
Time to mass conversion
  accurate mass, 55–57
  nominal mass, 47–51
Tomatidine, 72
Total ion current (TIC)
  chromatogram, 80
  percentage accounted for, 190, 195, 196, 202
  sampling of signal, 23
Training set, 164
Transformations, use of autoscaling after, 173–174

Tymnet, 3, 213
Typical factors, 182

U

Unsupervised learning, 151, 179–182
Urine analysis, 82, 90, 142, 144

V

Vertex graphs, 207
Vibration, 59

W

Weak signals, detection by ion counting, 240
Weight vector, 157
Weighting, 107, 167
  variance based, 160
Wiley Registry, 103, 105, 136